职业技能英语系列教材

机电英语
（第二版）

English for Mechanical and Electrical Engineering
(Second Edition)

主　编　李玉萍
副主编　谷士艳
编　者　秦军伟　吴　娜　王　莉
　　　　汪海涛　孙　刚

北京大学出版社
PEKING UNIVERSITY PRESS

图书在版编目(CIP)数据

机电英语 / 李玉萍主编. —2版. —北京：北京大学出版社，2021.4
职业技能英语系列教材
ISBN 978-7-301-31991-8

Ⅰ.①机…　Ⅱ.①李…　Ⅲ.①机电工程—英语—高等职业教育—教材　Ⅳ.①TH

中国版本图书馆CIP数据核字(2021)第022838号

书　　　名	机电英语(第二版)	
	JIDIAN YINGYU (DI-ER BAN)	
著作责任者	李玉萍　主编	
责任编辑	李　颖	
标准书号	ISBN 978-7-301-31991-8	
出版发行	北京大学出版社	
地　　　址	北京市海淀区成府路205号　100871	
网　　　址	http://www.pup.cn　　新浪微博:@北京大学出版社	
电子信箱	evalee1770@sina.com	
电　　　话	邮购部 010-62752015　发行部 010-62750672　编辑部 010-62754382	
印　刷　者	三河市博文印刷有限公司	
经　销　者	新华书店	

787毫米×1092毫米　16开本　11.5印张　317千字
2008年6月第1版
2021年4月第2版　2023年8月第2次印刷

定　　　价　45.00元

总　序

　　我国高职高专教育的春天来到了。随着国家对高职高专教育重视程度的加深，职业技能教材体系的建设成为了当务之急。高职高专过去沿用和压缩大学本科教材的时代一去不复返了。

　　语言学家 Harmer 指出："如果我们希望学生学到的语言是在真实生活中能够使用的语言，那么在教材编写中接受技能和产出技能的培养也应该像在生活中那样有机地结合在一起。"

　　教改的关键在教师，教师的关键在教材，教材的关键在理念。我们依据《高职高专教育英语课程教学基本要求》的精神和编者做了大量调查，兼承"实用为主，够用为度，学以致用，触类旁通"的原则，历经两年艰辛，为高职高专学生编写了这套专业技能课和实训课的英语教材。

　　本套教材的内容贴近工作岗位，突出岗位情景英语，是一套职场英语教材，具有很强的实用性、仿真性、职业性，其特色体现在以下几个方面：

　　1. 开放性

　　本套教材在坚持编写理念、原则及体例的前提下，不断增加新的行业或岗位技能英语分册作为教材的延续。

　　2. 国际性

　　本套教材以国内自编为主，以国外引进为辅，取长补短，浑然一体。目前已从德国引进了某些行业的技能英语教材，还将从德国或他国引进优秀教材经过本土化后奉献给广大师生。

　　3. 职业性

　　本套教材是由高职院校教师与行业专家针对具体工作岗位、情景过程共同设计编写。同时注重与行业资格证书相结合。

　　4. 任务性

　　基于完成某岗位工作任务而需要的英语知识和技能是本套教材的由来与初衷。因此，各分册均以任务型练习为主。

5. 实用性

本套教材注重基础词汇的复习和专业词汇的补充。适合在校最后一学期的英语教学，着重培养和训练学生初步具有与其日后职业生涯所必需的英语交际能力。

本套教材在编写过程中，参考和引用了国内外作者的相关资料，得到了北京大学出版社外语编辑部的倾力相助，在此，一并向他们表示敬意和感谢。由于本套教材是一种创新和尝试，书中瑕疵必定不少，敬请指正。

丁国声

教育部高职高专英语类专业教学指导委员会委员

河北省高校外语教学研究会副会长

秦皇岛外国语职业学院院长

2008年6月

编写说明

　　本书是根据高等职业教育机电专业教学需求而编写的专业英语教材。本教材在编写中遵照高等职业教育的应用性特征,注重培养学生实际运用语言的能力,突出教学内容的实用性和针对性。

　　本教材共9个单元,选材涉及机械(机械零件,锻造、冲压和轧制,模具,机床、工程材料),计算机数控(计算机辅助设计、制造,计算机辅助工艺设计,计算机集成制造系统),电子与信息技术(集成电路)和应用技术(电火花加工,电动机,工业机器人,机电一体化)。每个单元由导入、课文阅读和课后阅读三部分组成,内容由浅入深,从基础知识到实际应用,语言规范,难度适中。既突出专业特色,又能充分体现英语教学的规律。

　　本教材在编写中注意以学生为中心,注重语言技能与职业知识技能的结合,强调职业仿真环境下工作语言情景的导入,图文并茂,力求增强教学的直观性,降低学习难度,增强趣味性和知识性。通过最新应用实例,体现教学内容的实时性,使学生在了解岗位主要流程、工作内容、工作职责、相关知识、文化背景和职业操守的同时,达到能运用英语自如应对涉外工作的目的。

　　本书在编写过程中得到了编者所在学校领导的支持,在此表示衷心的感谢。由于时间和水平有限,书中难免有缺陷与不足之处,欢迎广大读者不吝指正。

编者

2020.7

Contents

Machine Elements
机械零件

Introduction

The function of a mechanism is to transmit or transform from one rigid body to another as part of the action of a machine. There are three types of common mechanical devices that can be used as basic elements of a mechanism.

Machine Elements

1. Gear system, in which toothed members in contact transmit motion between rotating shafts.

2. Cam system, where a uniform motion of an input member is converted into a nonuniform motion of the output member.

3. Plane and spatial linkages are also useful in creating mechanical motions for a point or rigid body.

Gear

Gears are toothed wheels meshed together to transmit motion and force. In any pair of gears the larger one will rotate more slowly than the smaller one, but will

rotate with greater force. Each gear in a series reverses the direction of rotation of the previous gear.

You can have any numbers of gears connected together and they can be different in shapes and sizes. Each time you pass power from one gear wheel to another, you can do one of three things.

Gears

Increase speed: If you connect two gears together

and the first one has more teeth than the second one, the second one has to turn round much faster to keep up. So this arrangement means the second wheel turns faster than the first one but with less force.

Increase force: If the second wheel in a pair of gears has more teeth than the first one (that is, if it's a larger wheel), it turns slower than the first one but with more force. Turn the larger wheel and the smaller wheel goes slower but has more force.

Change direction: When two gears mesh together, the second one always turns in the opposite direction. So if the first one turns clockwise, the second one must turn counterclockwise. You can also use specially shaped gears to make the power of a machine turn through an angle. In a car, for example, the differential uses a cone-shaped bevel gear to turn the drive shaft's power through 90 degrees and turn the back wheels.

Cam

Cam

A cam is a machine element having a curved outline or a curved groove, which, by its oscillation or rotation motion, gives a predetermined specified motion to another element called the follower. Cam mechanism is a high-pair mechanism composed of cam, follower and rack. Cam mechanism is conveniently used to transform one of the simple motions, such as rotation, into any other motions, such as translation, oscillation.

Pulley

Pulley

A pulley is simply a collection of one or more wheels over which you loop a rope to make it easier to lift things. The more wheels you have, and the more times you loop the rope around them ,the more you can lift.

In engineering, the kind of pulley is sometimes called a block and tackle: the wheels and their mounts are the blocks and the ropes that loop around them are the tackle. One block is fixed at the top and the other block moves up with the load.

Block

Tackle

Block

Pulley Block

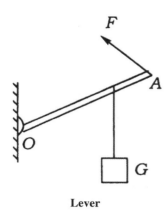

Lever

Screw

More generally, to engineers, a pulley is a wheel over which you loop a rope or a belt to connect one part of a machine to another, whether it's lifting things, transmitting power, or doing anything else. In simple science, though,we tend to use 'pulley' just to mean a bunch of wheels and ropes for lifting.

Lever

A lever is a stiff rod that rotates around a pivot point. Downward motion at one end results in upward motion at the other end. Depending on where the pivot point is located, a lever can multiply either the force applied or the distance over which the force is applied.

Screw

A screw is a central core with a thread or groove wrapped around to form a helix. While turning, a screw converts a rotary motion into a forward or backward motion.

1. The following are the elements of a machine. Do you know the meanings of them? Can you add any?

nut	spring	gear
cam	shaft	bearing
coupling	clutch	lever
hub	pulley	supporting structure
rivet	crankshaft	connecting rod
bolt	screw	fastener

The Elements of a Machine

Vocabulary Assistant

transmit	传输,转送	rigid	刚性的
gear	齿轮	shaft	轴
rotating shaft	转轴	cam	凸轮
uniform motion	均匀(等速)运动	convert	使转变,使……改变
plane	平面的	spatial linkage	空间联结
wheel	轮	mesh	啮合
reverse	颠倒,转动	tooth	轮齿;啮合
differential	差速器	bevel gear	锥齿轮
groove	槽纹,刻槽	oscillation	摆动,振动
predetermine	事先安排,预先设定	follower	从动轮
translation	移动,变换	pulley	滑车,滑轮
loop	使成环	rope	绳索
block	滑车	tackle	索具
belt	带	lever	杆,杠杆
pivot point	支点	screw	螺丝钉
helix	螺旋,螺旋状物		

2. Look at the following pictures. Do you know what they are? Then discuss the function of each with a partner.

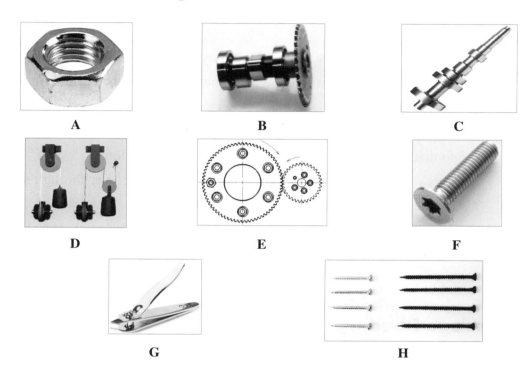

3. Look at the following elements of a machine. Can you give the English version to each of them?

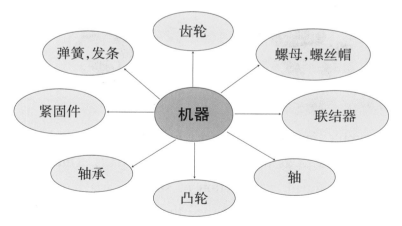

The Elements of a Machine

4. Match the definitions with the terms.

> A. gear   B. bearing   C. cam   D. shaft

_____ 1) This device can transmit exact motions at specific time in cycle. It has a curved grooved surface which mates with follower and imparts motion to it.

_____ 2) This device is wheel with teeth that engages or meshes with each other so that they work in pairs to transmit or change motion.

_____ 3) This device is mounted in bearings and transmits power through such devices as gears, pulleys, cams, and clutches.

_____ 4) This element is a support system for rotating, oscillating, or translating machine elements, in which friction has been greatly reduced.

Passage Reading

Machine Elements
机械零件

1 However simple, any machine is a combination of individual components generally referred to as machine elements or parts. Thus, if a machine is completely dismantled, a collection of simple parts remains such as nuts, bolts, springs, gears, cams and shafts—the building block of all machinery. A machine element is, therefore, a single unit designed to perform a specific function and capable of combining with other elements. Sometimes certain elements are associated in pairs, such as nuts and bolts or keys and shafts. In other instances, a group of elements is combined to form a subassembly,

such as bearings, couplings, and clutches.

> **Questions:** *What is a machine?*
>
> *Can you list some machine elements?*

2 The most common example of machine elements is gear, which is a combination of wheel and lever to form a toothed wheel. The rotation of this gear on a hub or shaft drives other gears that may rotate faster or slower, depending upon the number of teeth on the basic wheels.

3 Other fundamental machine elements include wheel and lever. A wheel must have a shaft on which it may rotate. The wheel is fastened to the shafts with couplings, and the shaft must rest in bearings. The supporting structure may be assembled with bolts, rivets or by means of welding. Proper application of these machine elements depends upon knowledge of the force on the structure and the strength of the materials employed.

> **Questions:** *What is a gear?*
>
> *What can influence the proper application of machine elements?*

4 The individual reliability of machine elements becomes the basis for estimating the overall life expectancy of a complete machine.

5 Many machine elements are thoroughly standardized. Testing and practical experience have established the most suitable dimensions for common structural and mechanical parts. Through standardization, uniformity of practice and resulting economics are obtained. Not all machine parts in use are standardized, however. In the automotive industry, only fasteners, bearings, bushings, chains, and belts are standardized. Crankshafts and connecting rods are not standardized.

> **Questions:** *How can we estimate the overall life expectancy of a complete machine?*
>
> *What can we obtain through standardization?*

Vocabulary Assistant

combination	结合,联合	component	成分,元件,部件
dismantle	拆除	capable	有可能的
combine	(使)联合	subassembly	部件,组件
rotate	旋转	fasten	结牢
assemble	装配	weld	焊接
application	应用	reliability	可靠性
overall	全部的,全体的	life expectancy	平均寿命,预期寿命
standardize	使标准化	dimension	大小,体积
uniformity	同样,一致	automotive	汽车的
automotive industry	汽车工业	bushing	[机]轴衬;[电工]套管
chain	链条,电路		

1. Fill in the table below by giving the corresponding translation.

English	Chinese
automotive	
	联结器
uniform motion	
	支承结构
crankshaft	
	螺钉
rotate	
	离合器
uniformity	
	寿命

2. Find the definition in Column B which matches the words in Column A.

A	B
_____ 1) component	a. tear down
_____ 2) motion	b. a long thin implement made of metal or wood
_____ 3) dismantle	c. link connecting two parts
_____ 4) collection	d. part of a machine element

_____ 5) rod   e. being basic

_____ 6) coupling   f. the condition of standards

_____ 7) fundamental   g. a change in the position or location of something

_____ 8) standardization   h. including everything; containing all

_____ 9) overall   i. put or add together

_____ 10) combine   j. several things grouped together or considered as a whole

3. Use the correct form of the words from the box to complete the sentences.

| transmit   machine   fasten   subassembly   bolt   standard |

1) Mechanism forms the basic geometrical elements of many _____ devices including automatic packaging machinery, typewriters, mechanical toys, textile machinery and others.

2) Gears are normally used for the _____ of motion with a constant angular velocity ratio, although noncircular gears can be used for nonuniform motion.

3) In the case of automobile, there are literally thousands of parts that are _____ together to produce the total product.

4) In the modern industrialized world, the wealth and living _____ of a nation are closely linked with their capability to design and manufacture engineering products.

5) _____, gears, and chains are the typical examples of the universal mechanical components.

6) A group of elements is combined to form a _____ , such as bearings, couplings, and clutches.

4. Pair Activities: Robert and Jake are students majoring in Mechanical and Electrical Engineering. They are talking about the elements of a machine. (Robert: R, Jake: J)

R: Do you know gear?

J: Yes, it's the most common example of machine elements.

R: Could you tell me the function of gear?

J: Gear is used to transmit power or motion from one shaft to another, or used to change the speed or the torque of one shaft with relation to another.

R: How is it used in our daily life?

J: There are many. One of the first mechanism invented using gears is clock. Besides, there are bikes, cars, elevators and so on.

R: Thank you. Then what about fastener? What is it?

J: The fastener permits one part to be joined to the other part and it is involved in almost any designs. There are three main classifications of fasteners, which are described as removable, semi-permanent, and permanent.

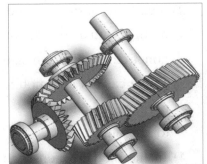

R: Then, are there any differences among these three types?

J: Yes, just as the terms indicate, the removable one permits the parts to be readily disconnected without damaging the fastener. An example is the ordinary nut-and-bolt fastener. Semi-permanent type can be disconnected, but some damage usually occurs to the fastener. One such example is a cotter pin. And when the permanent one is used, it is intended that the parts will never be disassembled. Examples of this type are riveted joints and welded points.

R: Yes, I see. Then what's the importance of fastener?

J: The importance of fastener can be realized when we refer to any complex product. In the case of the automobile, there are literally thousands of parts which are fastened together to produce the total product. The failure or loosening of a single fastener could

result in a simple nuisance such as a door rattle or in a serious situation such as a wheel coming off. Such possibilities must be taken into account in the selection of the type of fastener for the specific application.

R: That is really important. Then, how about bearing? Is it important too?

J: Sure. You know, bearing is commonly applied in the support of a rotating shaft that is transmitting power from one location to another. Since there is always relative motion between a bearing and its mating surface, there is friction. In many instances, such as the design of pulleys, brakes, and clutches, friction is desirable. However, in the case of bearing, the reduction of friction is one of the prime considerations.

R: Why?

J: You should know friction results in loss of power, the generation of heat, and increased wear of mating surfaces.

R: Then, how about shaft?

J: Nearly all machines contain shaft.

R: So it is equally important?

J: Sure. You know the most common shape for shaft is circular and the cross section can be either solid or hollow.

R: How is it applied in transmitting power?

J: Shafts are mounted in bearings and transmit power through such devices as gears, pulleys, cams and clutches. These devices introduce forces which attempt to bend the shaft; therefore, the shaft must be rigid enough to prevent overloading of the supporting bearings.

R: That is really great! Thanks a lot.

J: My pleasure.

Answer the following questions:

1) What is the function of gear?

2) What is fastener?

3) How is shaft used?

4) What is the importance of bearing?

5. Application

Machine elements are commonly used in all fields of industrial manufacturing. The mechanical products are always a combination of various machine elements. Work with a partner to discuss what kind of machine elements are used in the following products. And how are they used?

A

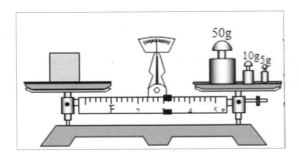

B

C

D

Vocabulary Assistant

torque 扭矩,转矩,力矩

classification 分类

permanent 永久的

refer to 参考,涉及,关于,查阅

nuisance 令人讨厌的事物

prime 首要的

steering wheel 方向盘,转向轮

involve 与……有关系;涉及

semi-permanent 半永久的,暂时的

cotter pin 开口销

complex 复杂的

rattle 一连串短而尖锐的声音

spring balance 弹簧秤

combine harvester 联合收割机

Over to you:

True or False Questions

_____ 1) Machine is a single unit designed to perform a specific function and capable of combining with other elements.

_____ 2) All of the elements of a machine are associated in pairs.

_____ 3) The rotation of the gear depends upon the number of teeth on the basic wheels.

_____ 4) Proper application of the machine elements depends upon knowledge of the force on the structure and the strength of the materials employed.

_____ 5) Machine parts, such as fasteners, bearings, bushings, chains, and belts, are standardized in the automotive industry only.

Further Reading

Design of Machine Elements
机械零件的设计

Design is one of the most important engineering functions for it is through design that new products and processes are born and that old ones are improved. Machinery design is an important technological basic course in mechanical engineering education. Its objective is to provide the concepts, procedures, data, and decision analysis techniques necessary to design machine elements commonly found in mechanical devices and systems and to develop engineering students' competence of machine design that is the primary concern of machinery manufacturing and the key to manufacturing good products.

Design requires a breadth of knowledge extending over many areas, and a sound analytic ability. It requires an ability to recognize the phenomena involved and to synthesize an integrated solution. Design requires sound engineering judgment as well as a good grasp of the underlying basic science and mathematics.

The principles of design are of course universal. The same theory or equations may be applied to a very small part, as in an instrument, or to a larger but similar part used in a piece of heavy equipment. In no case, however, should mathematical calculation be looked upon as absolute and final. They are all subject to the accuracy of the various assumptions which must necessarily be made in engineering work. Sometimes, only portions of the total number of parts in a machine are designed on the basis of analytic calculations. The form and size of the remaining parts are then usually determined by practical considerations. On the other hand, if the machine is very expensive, or if weight is factor, as in airplanes, design computations may then be made for almost all the parts.

The purpose of the design calculations is to predict the stress or deformation in the part. So it may safely carry the loads, and it may last for the expected life of the machine. All calculations depend on the physical properties of the construction materials as determined by laboratory tests. A rational method of design is to take the results of relatively simple and fundamental tests and apply them to all the complicated and involved situations encountered in today's machinery.

In addition, it has been proved that such details as surface condition, manufacturing tol-

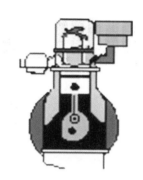

erances, and heat treatment have marked effect on the strength and useful life of a machine part. The design and drafting departments must specify completely all such particularities, and thus exercise the necessary close control over the finished product.

Training in rapid and accurate numerical work is invaluable to the designer. The designer should keep an accurate note book, as it is necessary for him to refer to work which he has done in the past. A sketch, carefully drawn to scale, is also necessary, and provides a convenient place for putting down a portion of the data used in connection with the problem. It goes without saying that all data, assumptions, equations, and calculations should be written down in full in order to be intelligible when referred to later.

Vocabulary Assistant

analytic 分析的	synthesize 合成,结合
equation 相等,平衡,方程式	calculation 计算,考虑
assumption 假设	computation 计算,估计
stress 压力	deformation 变形
rational 合理的	manufacturing tolerance 制造公差
tolerance 公差	drafting 起草
specify 列举,规定	particularity 特性,特质
finished product 成品	invaluable 无价的,非常宝贵的
sketch 草图,设计图	intelligible 可理解的

Over to you:

1) What are the principles of designing machine elements?

2) What is the purpose of design calculations?

Unit 2

Metal Forming Processes
金属成形过程

Introduction

Bulk deformation of metals refers to various processes, such as forging, stamping, rolling, casting, and welding, where there is a controlled plastic flow or working of metals into useful shapes.

Forging is an important hot-forming process. It is used in producing components of all shapes and sizes, from quite small items to large units weighing several tons.

Forging

Forging is the process by which metal is heated and is shaped by plastic deformation by suitably applying compressive force. Usually the compressive force is in the form of hammer blows using a power hammer or a press.

The forging family includes: hammer forging, press forging, open-die forging, impression-die forging and extrusion forging.

Stamping

Stamping is the process that metal in the form of hot-rolled or cold-rolled sheets or strips may be formed into many shapes by forcing the sheet into the impressions in metal dies. It involves many different operations, such as bending, drawing, punching, etc.

The equipments of stamping can be categorized into two types: mechanical presses and hydraulic presses.

Mechanical Presses: Mechanical presses has a mechanical flywheel to store the energy, and transfer it to

Rolling

the punch and to the workpiece. They range in size from 20 tons up to 6000 tons. Strokes range from 5 to 500 mm (0.2 to 20 in) and speeds from 20 to 1500 strokes per minute. Mechanical presses are well suited for high-speed blanking, shallow drawing and for making precision parts.

Hydraulic Presses: Hydraulic Presses use hydraulics to deliver a controlled force. Tonnage can vary from 20 to 10,000 tons. Strokes can vary from 10 to 800 mm (0.4 to 32 in). Hydraulic presses can deliver the full power at any point in the stroke; variable tonnage with overload protection; and adjustable stroke and speed. Hydraulic presses are suitable for deep-drawing, compound die action as in blanking with forming or coining, low speed high tonnage blanking, and force type of forming rather than displacement type of forming.

Rolling is the process of shaping metal in a machine called rolling mill. It is extensively used in the manufacturing of plates and sheets, structural beams, and so on.

Casting

Casting is a manufacturing process in which **molten** metal is **poured** or **injected** and allowed to **solidify** in a suitably shaped mold **cavity**. During or after cooling, the cast part is removed from the mold and then processed for delivery.

Casting processes and cast-material technologies vary from simple to highly complex. Material and process selection depends on the part's complexity and function, the product's quality specifications, and the projected cost level.

Castings are parts that are made close to their final dimensions by a casting process. With a history dating back 6,000 years, the various casting processes are in a state of continuous refinement and evolution as technological advances are being made.

Welding is the process of permanently **joining** two or more metal parts, by melting both materials. The molten materials quickly cool, and the two metals are permanently bonded. **Spot welding** and **seam welding** are two very popular methods used for sheet metal parts. The simplest method of welding two pieces of metal together is known as pressure welding. The ends of

Welding

metal are heated to a white heat—for iron, the welding temperature should be about 1300℃—in a flame. At this temperature the metal becomes plastic. The ends are then pressed or hammered together, and the joint is smoothed off.

Mould is a fundamental technological device for industrial production. It is the core part of manufacturing process because its cavity gives its shape. There are many kinds of mould, such as casting & forging dies, ceramic moulds, die-casting moulds, drawing dies, injection moulds, glass moulds, magnetic moulds, metal extruding moulds, plastic and rubber moulds, plastic extruding moulds, powder metallurgical moulds, compressing moulds, etc. The following is the introduction of some of the moulding processes and the corresponding moulds used.

▶ **Compression moulding** is the least complex of the moulding processes and is ideal for large parts or low-quantity production. It is often used for proto-typing where samples are needed for testing fit and form into assemblies. It is best suited for designs where tight tolerances are not required.

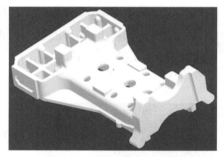

Mould

▶ **Injection moulding** is the most complex of the moulding processes. Its cycle time is much faster than other processes and the part cost can be low, particularly when the process is automated. It is well suited for moulding delicately shaped parts

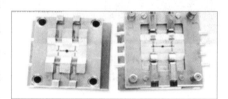

because high pressure is maintained on the material to push it into every corner of the mould cavity.

▶ **Extrusion moulding** is quite simple, but it requires great care in the setting up, manufacturing and final processing to ensure consistency of products. Most extrusion moulds are simply one round piece of steel with the profile of the intended extrusion wire cut into them. Allowances are made for the shrinkage, expansion of the intended compound. Extrusion dies are the least complex of the moulds.

1. Look at the following activities concerning metal forming process. Do you know what each of them refers to? Can you add any?

■ plastic working of metal

- drawing the dies together
- deep drawing
- removing the scale
- closed impression die
- drop forging
- hot rolling
- cold rolling
- kneading the metal
- heat treatment
- bending
- punching

Vocabulary Assistant

bulk　体积;批量

forging　锻造

rolling　碾,轧

plastic　塑性的

shape　形状

unit　部件;单元

force　力,力量

die　模型模具

extrusion　挤压

strip　条带

punch　冲孔,冲压

mechanical press 机械冲床

hydraulic press 液压机,水压机

workpiece　工件

tonnage　吨位

compound die 复式压模

plate　金属板

molten　熔化的

deformation　变形

stamping　冲压,模锻

casting　锻造

hot-forming　热成型

size　尺寸

compressive　压缩的

hammer　锤击

impression　压力,压痕

sheet　薄板

metal die 金属压型

categorize　分类

hydraulic　水力的,液压的

flywheel　飞轮

stoke　冲程

overload　过载

mill　轧钢机

structural beam　结构横梁

pour　倾倒,浇灌

inject 注入	solidify 使固化
cavity 型腔	join 密缝焊
spot welding 点焊	seam welding 缝焊
white heat 白热	flame 火焰
smooth 使光滑,使平坦	mould/mold 模具
forging die 锻模	ceramic mould 陶瓷铸型
die-casting mould 压铸模	drawing die 拉模
injection mould 注塑模	magnetic mould 磁铁成型模
extrude 挤压	
powder metallurgical mould 粉末冶金模	
compressing mould 冲压模,压塑模	proto-typing 初始制模
tolerance 公差	extrusion moulding 挤压成型模
consistency 一致,相等	profile 断面,剖面
allowance 公差;余量	shrinkage 收缩
expansion 膨胀	

Then answer the following questions:

1) *What is forging?*

2) *What is rolling?*

3) *What is stamping?*

4) *What is moulding? Is it connected with casting?*

2. Look at the following activities. Do you know what each of them refers to? Then work with a partner to discuss the process of hot forging.

Vocabulary Assistant

hot forging 热模锻	bar 条,棒,杆
blanking 下料,冲裁,切料	trimming 切边,修整
edge 镶边,飞边	heat treatment 热处理
correction 校正	inspection 检查
depositing 入库,沉淀,镀层	

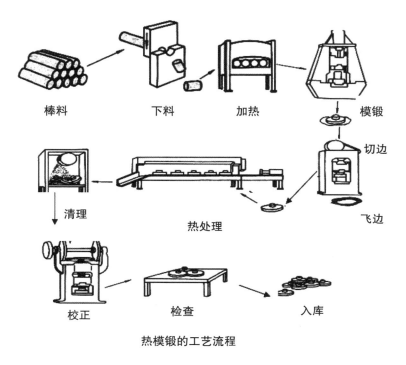

棒料　　　　下料　　　　加热　　　　模锻
　　　　　　　　　　　　　　　　　　　切边
清理　　　　　热处理　　　　　　　飞边
校正　　　　检查　　　　　入库

热模锻的工艺流程

3. Match the definitions with the terms.

A. knead	B. stamp	C. extrude	D. groove

_____ 1) to pound or crush with a heavy instrument

_____ 2) a long narrow channel or depression

_____ 3) to work and press into a mass with or as if with the hands

_____ 4) to force, press, or to shape (as metal or plastic) by forcing through a die

Passage Reading

Metal Forming Processes
金属成形过程

1 Metal forming is one of the fundamental manufacturing processes. It plays an important part in metallurgy, machine-building, power, automobile, railroad, aerospace, ship-building, weapon, chemical, electronics, instrument

and meter making and light industries. Metal forming is a plastic working of materials. In it, a simple part, or a sheet blank is plastically deformed between dies to obtain the desired final shape. Metal forming includes massive processes such as forging, stamping and rolling, etc.

2 In forging, a piece of metal, roughly or approximately of the desired shape, is placed between die faces having the exact form of the finished piece, and forced to take this form by drawing the dies together. This method is widely used for the manufacturing of parts both of steel and brass. Large ingots are now almost always forged with hydraulic presses instead of with steam hammers, since the work done by a press goes deeper. Few forgings of the types are produced without some of heat treatment. Untreated forging are usually relatively low-carbon steel parts for noncritical applications or are parts intended for further hot mechanical work and subsequent heat treatment.

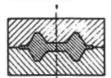   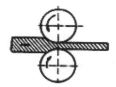

Drop Forging   **The Rolling of Plate**

Casting

Metal Forming Processes

Questions: *Where is metal forming used?*

What is the process of forging?

How is large ingot forged?

3 Metal in the form of hot-rolled or cold-rolled sheets or strips may be formed into many shapes by forcing the sheets into the impressions in metal dies. This process is known as stamping, or pressing of metals, and involves many different operations, such as bending, drawing, punching, etc. These operations are successfully carried out through the proper die design and proper operation of the press in which the dies are placed. Sheets or strips of steel, brass, copper, aluminum, or other

metal are placed between dies, and a slow action of the power press brings the dies together, forcing the metal to assume the required shape.

Questions: *How is metal formed into shapes?*

How is stamping carried out?

4 Rolling is the process of shaping metal in a machine called rolling mill. Ingots of metal are rolled by forcing them between two rollers rotating in opposite directions, thus pressing the metal into the required shape. If the rollers have no grooves on their surfaces, the metal is rolled into the form of a sheet or plate. If the rollers have grooves of a certain shape, the metal will take the form of these grooves, thus it may be shaped into the form of bars or rods. There are two kinds of rolling: hot rolling and cold rolling. Before cold rolling, the scale covering the surface of the hot-rolled object should be removed. Cold rolling produces a higher surface finishing sheet and gives it a very exact size. The process has innumerable advantages. Many shapes may be manufactured in quantities at a

rapid rate and at a relatively low cost.

5 Casting is a widely used method of producing metal products, particularly those which are intricate. This process consists of making molds, preparing and melting the metal, pouring the metal into the molds, and cleaning the castings. The casting process is basically simple. First, a cavity is formed in a mold. The shape of the cavity determines the shape of the casting. Liquid (molten) metal is poured into the mold, and then is allowed to cool and

become solid. After the metal has been solidified, the casting can be removed from the mold. The procedure can be repeated for production of duplicate parts. A given shape may be produced in quantities in the millions.

> **Questions:** *How is rolling carried out?*
> *What is the process of casting?*

6 Arc welding is the most widely used form of welding. The electrical supply is low voltage but high amperage and may be either alternating or direct. In many cases, the supply for welding operations is obtained from the secondary of a step-down transformer connected to the mains supply. If direct current is required, rectification of the secondary output is needed. Also, many portable welding units are available in which current is generated by direct-current compound-wound generator driven by a petrol engine. The latter are particularly useful for working on construction sites.

> **Questions:** *What is arc welding?*
> *Where is the electrical supply for welding operations obtained from?*

Vocabulary Assistant

metallurgy 冶金,冶金学		automobile 汽车
aerospace 航空		electronics 电子学
approximately 大约		brass 黄铜
ingot 锭铁,工业纯铁		steam hammer 蒸汽锤
low-carbon 低碳的,含碳量低的		subsequent 随后的,接着发生的
aluminum 铝		roller 辊子,滚筒
relatively 相对地		intricate 复杂的,复制的
duplicate 复制的,副本的		arc welding 电弧焊
voltage 电压;伏特数		amperage 安培数
alternating 交流(电)的		secondary 次级的;次级电路
step-down 低压		transformer 变压器
mains supply 干线供电,交流电源		direct current 直流电
rectification 整流		portable 轻便的,便携式的
generate 发生,产生		compound-wound 复式励磁的
generator 发电机,发生器		petrol engine 汽油(发动)机
construction 建筑		site 场所,地点

1. Fill in the table below by giving the corresponding translation.

English	Chinese
aluminum	
	落锤锻造
steam hammer	
	钢型,冲模
automobile	
	水压机
ceramic mould	
	压痕
fundamental	
	模腔

2. Find the definition in Column B which matches the words in Column A.

A	B
1) casting	a. an electrical device by which alternating current of one voltage is changed to another voltage
2) welding	b. motor that converts thermal energy to mechanical work
3) liquid	c. hollow container with a particular shape, into which a soft or liquid substance is poured to set or cool into that shape
4) transformer	d. object formed by a mold
5) engine	e. a copy that corresponds to an original exactly
6) current	f. a machine that produces electricity
7) generator	g. fastening two pieces of metal together by softening with heat and applying pressure
8) mold	h. substance in the fluid state of matter having no fixed shape but a fixed volume
9) duplicate	i. to become solid or make something solid
10) solidify	j. a flow of electricity through a conductor

3. Use the correct form of the words from the box to complete the following sentences.

relative	scale	plastic	operate	die	manufacture

1) The whole _____ should only take about ten minutes to perform.

2) Because of the development of technology, thousands of people lost their jobs in_____ industry.

3) _____ speaking, it's not important.

4) Before cold rolling, the _____ covering the surface of the hot-rolled object should be removed.

5) The material is melted and poured into a _____ cavity corresponding to the desired geometry.

6) Hot working is defined as _____ deforming the metallic material above the recrystallization temperature.

4. Pair Activities: Several students are visiting a factory. Philip is an engineer there. Now he is asked to explain the metal forming process to the students. (A: Philip B: student)

A: Good morning, everyone! Welcome to our factory. It's my pleasure to be your guide, so if you have any question, please go ahead.

B: Thank you! Oh, look! What's that? I am so interested in that operation. Could you please tell me?

A: Oh, that is forging. It is a process of giving metal increased utility by shaping it, refining it, and improving its mechanical properties through controlled plastic deformation under impact or pressure.

B: How will this process be carried out?

A: Forging, while producing desired shape, actually kneads the metal. This kneading is known as "hot working".

B: What is "hot working"?

A: Hot working is a process of plastic deformation, which is accomplished above the recrystallization temperature.

B: Does it work at high temperature?

A: The use of the term "hot working" usually implies that the material is heated—but not always. For example, the recrystallization of lead takes place at a very low temperature. By the above definition, lead forged at room temperature is being hot worked, too.

B: Could you tell me the principal methods of hot working?

A: Sure. They are hammering, pressing, rolling and extrusion. Hammer forging consists of striking

the hot metal with a large semiautomatic hammer. If no dies are involved, the forging will be dependent mainly on the skill of the operator. If closed or impressed dies are used, one blow is struck for each of several die cavities.

B: Then, is there cold working?

A: Yes, cold working is defined as plastic deformation below the recrystallization temperature.

B: Does cold working work at low temperature?

A: Not so often. Several common metal, or their alloys, have recrystallization temperatures in the range of 750℃ ~ 990℃ to improve properties, several of these alloys are strained and hardened by forging in the range of 550℃~700℃. Although this is truly, by definition, called "cold working", it is often referred to as "warm working".

B: Then, what is the difference between hot working and cold working?

A: When deforming a metal into a useful shape, we use hot working more often than cold working.

B: Why?

A: This is because hot working is defined as plastically deforming the metallic material above the recrystallization temperature. It's suited for forming large parts, since the metal has a low yield strength and high ductility at elevated temperature.

B: Oh, I see. Thank you very much!

A: You are welcome.

Answer the following questions:

1) *What is hot working?*

2) *What is cold working?*

3) *What is the difference between hot working and cold working?*

Vocabulary Assistant

impact 冲
recrystallization 再结晶
lead 铅
semiautomatic 半自动的
strike 打, 撞击
strain 拉紧
yield 屈从, 屈服
ductility 延展性, 柔韧性

accomplish 完成
imply 暗示
principal 主要的
blow 击打
alloy 合金
suited 适合的
strength 力
elevate 升高

5. Application

Look at the following figures. What does each of them refer to? Find a correct explanation for each of them.

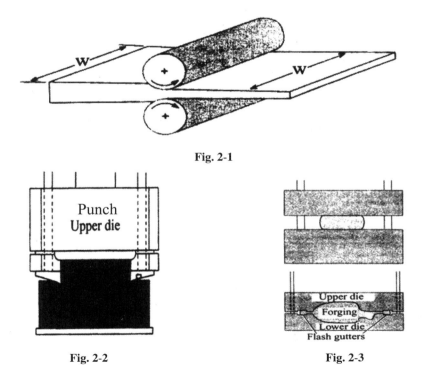

Fig. 2-1

Fig. 2-2   Fig. 2-3

_____ A. Rolling is used to produce metal plate and sheet.

_____ B. Forging deforms the material into a die cavity, producing relatively

complex shapes such as automotive crankshafts or connecting rods.

_____ C. Stamping is the process that metal in the form of hot-rolled or cold-rolled sheets or strips may be formed into many shapes by forcing the sheet into the impressions in metal dies. It involves many different operations, such as bending, drawing, punching, etc.

Over to you:

True or False Questions

_____ 1) The method of drop forging is widely used for the manufacturing of parts both of steel and brass.

_____ 2) Large ingots are now almost always forged with steam hammers instead of with hydraulic presses.

_____ 3) Forcing the sheet into the impressions in metal dies is known as stamping, or pressing of metals, and involves many different operations.

_____ 4) There are two kinds of rolling: hot rolling and cold rolling.

_____ 5) Many shapes may be manufactured in quantities at a rapid rate and at a relatively low cost by stamping.

Further Reading

Engineering Materials
工 程 材 料

Materials may be grouped in several ways. Scientists often classify materials by their state: solid, liquid, or gas. For industrial purpose, materials are divided into

engineering materials or nonengineering materials. Engineering materials are those used in manufacture and become parts of products, which may be further subdivided into metal, ceramic, polymer, etc.

Here, we just take metal and ceramic as examples.

Metals are elements that generally have good electrical and thermal conductivity. Many metals have high strength, high stiffness, and good ductility. Some metals, such as iron, cobalt and nickel, are magnetic. At extremely low temperature, some metals and intermetallic compounds become superconductors. Pure metals are elements which come from a particular area of the periodic table. Examples of pure metals include copper in electrical wires and aluminum in cooking foil or beverage cans. Alloys contain more than one metallic element. Their properties can be changed by changing the elements present in the alloy. Examples of metal alloys include stainless steel which is an alloy of iron, nickel, chromium; and gold jewelry which usually contains an alloy of gold and nickel.

Many metals and alloys have relatively high densities, which refer to the relation of mass to volume, and they often contain atoms with high atomic numbers, such as gold or lead. Some metal alloys like those based on aluminum, have low densities and are used in aerospace applications for fuel economy.

Many alloys also have high fracture toughness, which means they can withstand impact and they are durable. Fracture toughness can be described as a material's ability to avoid fracture, especially when a flaw is introduced.

Ceramics are broadly defined as any inorganic nonmetallic material. By this definition, ceramic materials would also include glass. A glass is an inorganic nonmetallic material that does not have a crystalline structure. Such materials are said to be amorphous.

Some of the useful properties of ceramics and glasses include high melting temperature, low density, high strength, stiffness, hardness, wear resistance, and corrosion resistance. Many ceramics are good electrical and thermal insulators. Some ceramics have special properties; some ceramics are magnetic materials; some are piezoelectric materials; and a few special ceramics are superconductors at very low temperatures. Ceramics and glasses have one major drawback: they are brittle.

Ceramics are not typically formed from the melt. This is because most ceramics will crack extensively upon cooling from the liquid state. Hence, all the simple and efficient manufacturing techniques used for glass production such as casting and blowing, which involve the molten state, cannot be used for the production of crystalline ceramics. Instead, "sintering" or "firing" is the process typically used. In sintering, ceramic powders are processed into compacted shapes and then heated to temperatures just below the melting point. At such temperatures, the powders react internally to remove porosity and fully dense articles can be obtained.

Vocabulary Assistant

engineering 工程(学)	subdivide 细分,再分
ceramic 陶瓷	polymer 聚合体
thermal conductivity 导热性	stiffness 坚硬,硬度
cobalt 钴	nickel 镍
intermetallic 金属间的	superconductor 超导体
periodic table (元素)周期表	copper 铜
foil 箔,金属薄片	beverage 饮料
stainless 不锈的	chromium 铬
atom 原子	density 密度
fracture 断裂	fracture toughness 断裂韧度
withstand 抵挡,经受	durable 耐用的
flaw 裂纹	inorganic 无机的
crystalline 水晶的	amorphous 无定形的
wear resistance 耐磨性	resistance 抵抗力
corrosion 侵蚀	thermal insulator 热绝缘体
piezoelectric 压电的	drawback 缺点
brittle 易碎的	crack 破裂,崩裂
blowing 吹炼	molten state 熔融状态
sintering 烧结	powder 粉末
compacted 紧密的,紧凑的	porosity 多孔性

Over to you:

    1. What are engineering materials?

    2. What are metals? And what are pure metals?

    3. What are ceramics? And what are the properties of ceramics?

    4. Why can't ceramics be formed from the melt?

Unit 3

Lathe
车 床

Introduction

The lathe is one of the most useful and versatile machines in the workshop, and is capable of carrying out a wide variety of machining operations. The main components of the lathe are the headstock and tailstock at opposite ends of a bed, and a tool-post between them which holds the cutting tool. The

tool-post stands on a cross slide which enables it to move sidewards across the saddle or carriage as well as along it, depending on the tool-post.

Machines using basically the single-point cutting tools include: engine lathes, turret lathes, tracing and duplicating lathes, single-spindle automatic lathes, multi-spindle automatic lathes, shapers and planers, gear-cutting machines. There are

The Engine Lathe

many machines using multipoint cutting tools, such as: drilling machines, milling machines, broaching machines, sawing machines, gear-cutting machines.

The engine lathe is the basic turning machine from which other turning machines have been developed. The drive motor is located in the base and drives the spindle through a combination of belts and mounted between heavy-duty bearings, with the forward end used for mounting a drive plate to impart positive motion to the workpiece. It can accommodate only one tool at a time on the tool-post.

Turret lathe is basically an engine lathe with certain additional features to provide for semiautomatic operation and to reduce the opportunity for human error. It is capable of holding five or more tools on the revolving turret. The carriage of the turret lathe is provided

Turret Lathe

with T-slots for mounting a tool-holding device on both sides of the lathe ways with tools properly set for cutting when rotated into position. The carriage is also equipped with automatic stops that control the tool travel and provide good reproduction of cuts. The tailstock of the turret lathe is of hexagonal design, in which six tools can be mounted. Although a large amount of time is consumed in setting up the tools and stops for operation, the turret lathe, once set, can continue to duplicate operations with a minimum of operator skill until the tools become dulled and need replacing.

Tracing and Duplicating Lathe

Tracing and duplicating lathes are equipped with a duplicating device to automatically control the longitudinal and cross feed motions of the single-point cutting tool and provide a finished part of required shape and size in one or two passes of the tools.

Single-spindle automatic lathe uses a vertical turret as well as two cross slides. The work is fed through the machine spindle into the chuck, and the tools are operated automatically by cams.

Single-spindle Automatic Lathe

1. The following are the terms related with lathe. Do you know the meaning of each? Can you add any?

◆ Hand tools

◆ Vertical planing machines

◆ Vertical lathes

◆ CNC boring machines

◆ CNC shearing machines

◆ CNC drilling machines

◆ CNC lathes

◆ Cutting-off machines

◆ CNC EDM wire-cutting machines

◆ CNC electric discharge machines

◆ CNC machine tool fittings

◆ CNC milling machines

◆ CNC wire-cutting machines

◆ CNC grinding machines

Vocabulary Assistant

lathe 车床

workshop 车间,工厂

headstock 床头箱,主轴箱

tool-post 刀座,刀架

cross slide 横向滑板,横滑台

carriage 托架,滑动架

turret 六角刀架,转塔刀架

tracing and duplicating lathe 仿形车床

single-spindle automatic lathe 单轴自动车床

multi-spindle automatic lathe 多轴自动车床

shaper 牛头刨床

drilling machine 钻床,钻孔机

broaching machine 拉床,铰孔床

heavy-duty 耐受力强的,重型的

hexagonal 六角形的

longitudinal 纵向的

*hand tool 手工具

boring machine 镗床

shearing machine 剪切机

versatile 多技能的,多才多艺的

machining operation 机械加工

tailstock 尾架,尾座,顶针座

cutting tool 切割具,刀具

saddle 鞍座,管托,座架

engine lathe 普通车床

turret lathe 六角(转塔)车床

planer 龙门刨床

milling machine 铣床

mount 安装

revolve 旋转

minimum 最小值

vertical 立式的

wire-cutting 线切削

tool fitting 配件

grinding machine 磨床

---

*生词在习题中出现时,未标记颜色。

Answer the following questions:

1) What is lathe?

2) What are the main components of lathe?

3) What is the difference between the ordinary centre lathe and the turret lathe?

4) What kinds of lathes use the single-point cutting tools? And what use the multi-point ones?

2. **The following is a figure of the basic components of a lathe. Can you tell the name of each part? Then work with a partner to discuss the fanction of each part.**

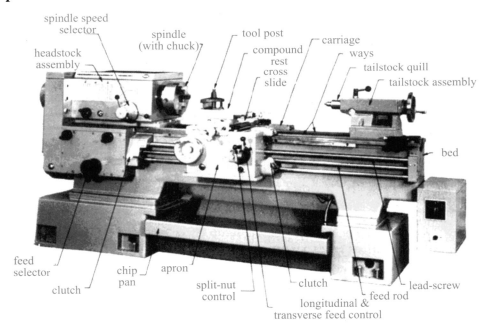

Basic Components of a Lathe

Vocabulary Assistant

headstock assembly　主轴箱组件　　　spindle　心轴,定轴

chuck　卡盘

compound rest　复式刀架,(车床的)小刀架

slide　滑动,滑板　　　　　　　　　quill　套筒轴

> tailstock assembly　尾座组件　　　bed　（机床）床身
> lead-screw　丝杠,导向螺杆　　　feed-rod　进刀杆
> apron　溜板,挡板,皮圈　　　　feed selector　进刀选择器

3. Find the definition in Column B which matches the words in Column A.

A	B
_____ 1) headstock	a. the backbone of a lathe
_____ 2) spindle	b. the part that can provide the means of rotating the work at various speeds
_____ 3) tool-post	c. the thing that can provide means for mounting and moving cutting tools
_____ 4) lead-screw	d. the connecting shaft between the feed box and apron
_____ 5) tailstock	e. The thing that is used to provide the accurate lead necessary for the thread cutting
_____ 6) carriage	f. sturdy heavy hollow shaft, mounted between heavy-duty bearings
_____ 7) bed	g. the thing that provides tool motion for accurate workpiece and good surface finishes
_____ 8) feed-rod	h. The tool that hangs from the front of the carriage and contains the driving gears
_____ 9) apron	i. hollow steel cylinder that can be moved several inches longitudinally
_____ 10) quill	j. H-shaped casting on which the cutting tool is mounted in a tool holder

4. Match the definitions with the terms.

> A. machine tool　　B. chuck　　C. lathe　　D. feed

_____ 1) Machines for cutting metals, such as lathes, drilling machines and milling machines

_____ 2) The movement of the cutter

      3) A machine tool for cutting metal from the surface of a round work fastened between the two lathe centers and turning around its axis

      4) Mounted on the headstock spindle, it is used to clamp the work so that it will rotate without wobbling while turning

Passage Reading

Lathe
车 床

1 Machine tools are machines for cutting metals. The most important machine tools used in industry are lathes, drilling machines and milling machines. Other kinds of metal working machines are not so widely used in machining metals as these three kinds.

2 A lathe is a machine tool for cutting metal from the surface of a round work fastened between the two lathe centers and turning around its axis. In turning the work a cutter moves in the direction parallel to the axis of rotation of the work or at an angle to this axis, cutting off the metal from the surface of the work. This movement of the cutter is called the feed. The cutter is clamped in the tool post that is mounted on the carriage. The lathe may feed the cutter by hand or may make it be fed automatically by means of special gears.

Questions: What are the most important machine tools in industry?
How does a lathe work?

3 The largest part of the lathe is called the bed on which the headstock and tailstock are fastened at opposite ends. On the upper part of the bed there are special ways upon which the carriage and the tailstock slide.

4 The two lathe centers are mounted in two spindles: one (the live center) is held in the headstock spindle while the other (dead center) in the tailstock spindle.

5 The lathe chuck is used for chucking the work, which is for clamping it so that it will rotate without wobbling while turning. The chuck, usually mounted on the headstock spindle, may be chucked in the so-called three-jaw universal chuck, all the jaws of which are moved to the center by turning the screw. But if the work is not perfectly round, the four-jaw independent chuck should be used.

> **Questions:** *What's the largest part of the lathe? And what is it used for?*
> *What is the lathe chuck? And how does it work?*
> *If the work is not perfectly round, what should be done?*

6 In turning different materials and works of different diameter, Lathes must be run at different speed. The gearbox contained in the headstock makes it possible to run the lathe at various speeds.

7 Before turning a work in the lathe, the lathe centers are to be aligned; that means that the axes of both centers must be on one line.

8 The alignment of the lathe centers may be tested by taking a cut and then measuring both ends of the cut with a micrometer.

> **Question:** *What makes it possible for a lathe to run at a different speed?*

9 Not all works should be fastened between the two centers of the lathe. A short work may be turned without using the dead center, by simply chucking it properly at the middle of the headstock.

Vocabulary Assistant

axis 轴	cutter 刀具
rotation 旋转	angle 角度
feed 进刀,走刀	clamp 夹,夹紧,固定
chuck 卡盘,夹具	wobble 摇摆
jaw 钳夹	diameter 直径
gearbox 变速箱	alignment 并列,定心,定位,调整
micrometer 千分尺	

1. Fill in the table below by giving the corresponding translation.

English	Chinese
cutter	
	主轴箱组件
wobble	
	复式刀架
diameter	
	旋转
tailstock	
	中拖板
jaw	
	配件

2. Use the correct form of the words from the box to complete the sentences.

cut   slide   motion   screw   mount   clamp

1) A machine tool provides the means for _____ tools to shape a workpiece to required dimensions.

2) The lathe bed is cast iron and provides accurately ground _____ surfaces on which the carriage rides.

3) The feed rod is used to provide tool _____ essential for accurate workpiece and good surface finishes.

4) Many lathe manufacturers combine the feed rod and the lead _____ in one, a practice that reduces the cost of the machine at the expense of accuracy.

5) The two lathe centers are _____ in two spindles: one (the live center) is held in the headstock spindle while the other (dead center) in the tailstock spindle.

6) The cutter is _____ in the tool post that is mounted on the carriage.

3. Match the items listed in the following two columns.

_____ 1) chuck   a. case that encloses a vehicle's gear mechanism

_____ 2) parallel   b. to clamp the work

_____ 3) gearbox   c. thing made up of two or more separate things combined together

_____ 4) compound   d. of, done with or controlled by the hands

_____ 5) manual   e. lines having the same distance between each other at every point

_____ 6) axis   f. line round which a turning object spins

4. Pair Activities: Jack is a student majoring in Mechanical and Electrical Engineering. After practicing in workshop, he is asked to explain lathe to the new student. (A: new student, B: Jack)

A:　What are the lathes used for?

B:　Lathes are the basic machines that are designed primarily to do turning, facing and boring.

A:　Then, what are the major functions a lathe performs?

B:　There are three major functions: first, it can rigidly support the workpiece or its holder and the cutting tool. Second, it provides relative motion between the workpiece and the cutting tools. Third, it provides a range of feeds and speeds.

A:　And could you explain the essential components of a lathe?

B:　Yes, you can look at the following diagram. The basic components of a lathe are the bed, headstock assembly, tailstock assembly, carriage

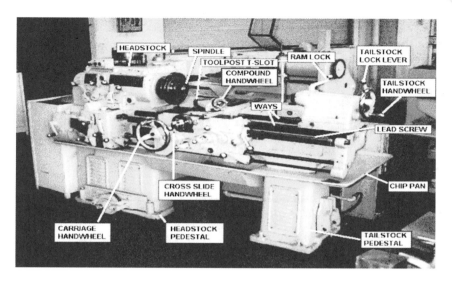

assembly, lead screw, tool-post, slide and spindle.

A: Then how does it work?

B: The lathe works like this: it can remove material by rotating the workpiece against a cutter to produce external or internal cylindrical or conical surfaces. It is also commonly used for the production of flat surfaces by facing.

A: Is the engine lathe the simplest one?

B: It's not the simplest one, but the basic turning machine from which other turning machines have been developed. For example, the turret lathe is basically an engine lathe with certain additional features to provide for semiautomatic operation and to reduce the opportunity for human error. The carriage of the turret lathe is provided with T-slots for mounting a tool-holding device on both sides of the lathe ways with tools properly set for cutting when rotated into position. Its carriage is equipped with automatic stops and tailstock is of hexagonal design. It is economically feasible only for production work.

A: I see. Thank you so much.

B: My pleasure.

Vocabulary Assistant

boring 钻孔

diagram 图表

cylindrical 圆柱的

feasible 可行的

holder 固定器,支持物,支架

assembly 装配,组件

conical 圆锥形的,圆锥的

Answer the following questions:

1) What are the major functions that a lathe performs?

2) How does a lathe work?

3) What is turret lathe?

5. Application

An engine lathe is a horizontally shaped piece of machinery that is most of-

ten used to turn metal manually. By turning the metal and using special cutting tools, the engine lathe is capable of forming the metal into specific shapes. As its name implies, the engine lathe is often used to create metal pieces for use in an engine, whether it be for an automobile, a tractor, a boat, or any other motorized vehicle or machine.

Although people use the engine lathe primarily for spinning sheet metals, it is also used for drilling, making square blocks, and creating shafts. Most modern tools were made with the help of an engine lathe. In addition, those who own an engine lathe can make their own tools with the machine. For these reasons, the engine lathe is often referred to as a reproductive machine.

The features of an engine lathe include gears, a carriage, a tailstock, and a stepped pulley used for various spindle speeds. The gears in the engine lathe are used to power the carriage. In turn, the carriage supports the cutting tools. The tailstock is used to support the hole-drilling process that takes place in the spindle.

A CNC lathe is a machine tool designed to remove material from a rotating workpiece, using a cutting tool. Some lathes can form hollow parts by a process called metal spinning. These parts have circular cross-sections. Metal and other materials can be turned on a lathe, including wood and plastics. CNC controlled lathes use a computer to control the process of making each part with repeated accuracy and precision.

Look at the following pictures. Do you know what each of them refers to? Find correct explanation for each of them.

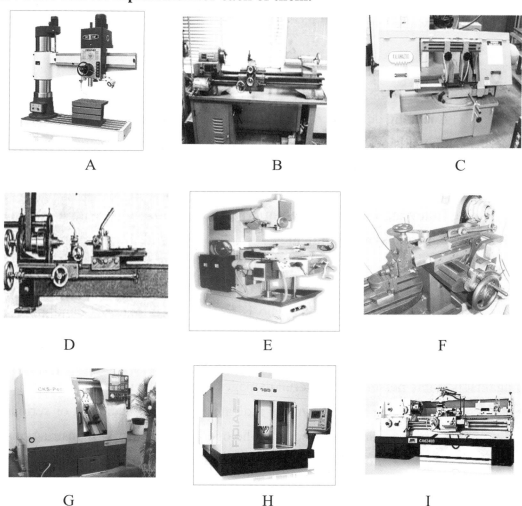

A	B	C
D	E	F
G	H	I

1) Drilling machine

2) Turret lathe

3) Horizontal Band Saw

4) Lathe

5) Milling Machine

6) High speed milling machine

7) CNC lathe

8) Shaping Machine

9) Horizontal lathe

10) CNC vertical lathe

11) Heavy-duty and large-size lathe

J　　　　　　　K

Over to you:

True or False Questions

____ 1) A locking device permits clamping the quill in any desired position.

____ 2) The tailstock provides the means for mounting and moving cutting tools.

____ 3) On most lathes the tool-post is actually mounted on a compound rest.

The following is a part of an operating manual. Work with a partner to discuss what each symbol indicates.

Safety Precautions

For safety reasons, anybody who is working with material-handling equipment must read, understand and observe these operating instructions. The following symbols and comments are used in the operating instructions. They warn against possible personal or material damages or assist the user.

 This symbol indicates that lethal accidents or serious injuries may occur if the operating and working instructions are not followed properly. Such warnings must be strictly kept to avoid lethal or serious injuries.

 This symbol warns against dangerous voltage! Never touch any live parts. Immediate death might be the consequence.

Warnings must be strictly observed to avoid lethal or serious injuries.

This symbol and the relevant comments draw your attention to something particular. Your work will be more efficient if you observe this information. Such information makes work easier or explains complex facts.

This symbol indicates prohibitive actions that must not be performed by the operator.

This symbol indicates compulsory action which must be performed by the operator.

This symbol informs the user that the system or its components may be damaged or other material damages may occur if the working and operating instructions are not observed exactly or not observed at all. Such instructions must be strictly observed to avoid material damages.

Further Reading

How to Use a Lathe
如何使用机床

The lathe is a machine tool used principally for shaping pieces of metal (and sometimes wood or other materials) by causing the workpiece to be held and rotated

by the lathe while a tool bit is advanced into the work causing the cutting action. The basic lathe that was designed to cut cylindrical metal stock has been developed further to produce screw threads, drilled holes, knurled surfaces, and crankshafts. Modern lathes offer a variety of rotating speeds and a means to manually and

automatically move the cutting tool into the workpiece. Machinists and maintenance shop personnel must be thoroughly familiar with the lathe and its operations to accomplish the repair and fabrication of needed parts.

Types of Lathes

Lathes can be divided into three types for easy identification: engine lathe, turret lathe, and special purpose lathes. Some smaller ones are bench-mounted and semi-portable. The larger lathes are floor-mounted and may require special transportation if they must be moved. Field and maintenance shops generally use a lathe that can be adapted to many operations and that is not too large to be moved from one work site to another. A trained operator can accomplish more machining jobs with the engine lathe than with any other machine tool. Turret lathes and special purpose lathes are usually used in production or job shops for mass production or specialized parts, while basic engine lathes are usually used for any type of lathe work. Further reference to lathes in this chapter will be about the various engine lathes.

The size of an engine lathe is determined by the largest piece of stock that can be machined. Before machining a workpiece, the following measurements must be considered: the diameter of the work that will swing over the bed and the length between lathe centers.

Slight differences in the various engine lathes make it easy to group them into three categories: lightweight bench engine lathes, precision tool room lathes, and gap lathes, which are also known as extension-type lathes.

Lightweight bench engine lathes are generally small lathes with a swing of 10 inches or less, mounted to a bench or table top. These lathes can accomplish most machining jobs, but may be limited due to the size of the material that can be turned.

Precision tool room lathes are also known as standard manufacturing lathes and are used for all lathe operations, such as turning, boring, drilling, reaming, produc-ing screw threads, taper turning, knurling, and radius forming, and can be adapted for special milling operations with the appropriate fixture. This type of lathe can handle workpieces up to 25 inches in diameter and up to 200 inches long. However, the general size is about a 15-inch swing with 36 to 48 inches between centers. Many tool room lathes are used for special tool and die production due to the high accuracy of the machine.

Gap or extension-type lathes are similar to tool room lathes except that gap lathes can be adjusted to machine larger diameter and longer workpieces. The operator can increase the swing by moving the bed a distance from the headstock, which is usually one or two feet. By sliding the bed away from the headstock, the gap lathe can be used to turn very long workpieces between centers.

Engine lathes all have the same general functional parts; even though the specific location or shape of a certain part may differ from one manufacturer. The bed is the foundation of the working parts of the lathe to another.

Care and Maintenance of Lathes

Lathes are highly accurate machine tools designed to operate around the clock if properly operated and maintained. Lathes must be lubricated and checked for adjustment before operation. Improper lubrication or loose nuts and bolts can cause excessive wear and dangerous operating conditions.

The lathe ways are precision ground surfaces and must not be used as tables for other tools and should be kept clean of grit and dirt. The lead screw and gears should be checked frequently for any metal chips that could be lodged in the gearing mechanisms. Check each lathe prior to operation for any missing parts or broken shear pins. Refer to the operator's instructions before attempting to lift any lathe. Newly installed lathes or lathes that are transported in mobile vehicles should be properly leveled before any operation to prevent vibration and wobble. Any lathes that are transported out of a normal shop environment should be protected from dust, excessive heat, and very cold conditions. Change the lubricant frequently if working in dusty conditions. In hot working areas, avoid overheating the motor or damaging any seals. Operate the lathe at slower speeds than normal when working in cold environments.

Safety

All lathe operators must be constantly aware of the safety hazards that are associated with using the lathe and must know all safety precautions to avoid accidents and injuries. Carelessness and ignorance are two great menaces to personal safety. Other hazards can be mechanically related to working with the lathe, such as proper machine maintenance and setup. Some important safety precautions to follow when using lathes are:

- Correct dress is important, remove rings and watches, roll sleeves above elbows.
- Always stop the lathe before making adjustments.
- Do not change spindle speeds until the lathe comes to a complete stop.
- Handle sharp cutters, centers, and drills with care.
- Remove chuck keys and wrenches before operating

- Always wear protective eye protection.
- Handle heavy chucks with care and protect the lathe ways with a block of wood when installing a chuck.
- Know where the emergency stop is before operating the lathe.
- Use pliers or a brush to remove chips and swarf, never your hands.
- Never lean on the lathe.
- Never lay tools directly on the lathe ways. If a separate table is not available, use a wide board with a cleat on each side to lay on the ways.
- Keep tools overhang as short as possible.
- Never attempt to measure work while it is turning.
- Never file lathe work unless the file has a handle.
- File left-handed if possible.
- Protect the lathe ways when grinding or filing.
- Use two hands when sanding the workpiece. Do not wrap sand paper or cloth around the workpiece.

Vocabulary Assistant

precaution	预防,预防措施	lethal	致命的
prohibitive	禁止的	compulsory	强制的,强迫的
tool bit	刀头,刀片	knurl	压花,压迫
maintenance	维护	ream	铰孔,扩孔
radius	半径;辐射线	gap	凹口,间隙,开口
lubricate	使润滑	grit	粗砂,沙砾
prior to	在……之前	vibration	颤抖,振动
wobble	摇摆,晃动	lubricant	润滑剂,润滑油
menace	威胁,恐吓	setup	装备
emergency	突发事件	plier	钳子,镊子
swarf	金属屑	cleat	夹板,楔子
file	锉		

Over to you:

1) What are lightweight bench engine lathes?

2) What are precision tool room lathes?

3) What precautions should we take when using lathes?

Unit 4

计算机辅助设计
AutoCAD

CAD / CAM / CAPP/ CNC
计算机辅助设计/制造/工艺设计/计算机数控

Introduction

Computer-aided design (CAD) is a computer technology that designs a product and documents the design's process. CAD may facilitate the manufacturing process by transferring detailed diagrams of a product's materials, processes, tolerances and dimensions with specific conventions for the product in question. It can be used as follows: (1) To produce detailed engineering designs through 3-D and 2-D drawings of the physical components of manufactured products. (2) To create conceptual design, product layout, strength and dynamic analysis of assembly and the manufacturing

computer-aided design

processes themselves. (3) To prepare environmental impact reports, in which computer-aided designs are used in photographs to produce a rendering of the appearance when the new structures are built.

Computer-aided manufacturing (CAM) can be defined as the use of computer systems to plan, manage and control the operation of a manufacturing plant through either direct or indirect interface with the plant's production resources. The applications of CAM fall into two broad categories: computer monitoring and control, manufacturing support applications.

Computer-aided design and computer-aided manufacturing are often combined into CAD/CAM systems. This combination allows the transfer of information from

Computer-aided process planning

the design stage into the stage of planning for the manufacture of a product, without the need to reenter the data on part geometry manually.

Computer-Aided Process Planning (CAPP) is a highly effective technology for discrete manufacturers with a significant number of products and process steps. Rapid strides are being made to develop generative planning capabilities and incorporate CAPP into a computer-integrated manufacturing architecture. As a result, many companies can achieve the benefits of CAPP with minimal cost and risk. Effective use of these tools can improve a manufacturer's competitive advantage.

Computer numerical control (CNC) is the key technology driving today's manufacturing tools and processes. Today, CNC machines are controlled directly from files created by CAM software packages, so that a part or assembly can go directly from design to manufacturing without the need of producing a drafted paper-drawing of the manufactured component.

CNC system is the base of modern digital and intelligent manufacturing technology. CNC does numerically directed interpolation of a cutting tool in the work envelop of a machine. The operating parameters of the CNC can be changed by a software load program. Computer numerical control (CNC) system is a special computer system that is equipped with certain interface circuits and servo drivers, and can do part or all the works an NC system do by running the software stored in its memories. Today, CNC machine tools are widely used in manufacturing enterprises.

CNC machine tool

1. **The following are the terms related with CAD, CAM, CAPP and CNC. Do you know the meaning of them? Can you add any?**

- computer monitoring and control
- manufacturing support application
- peripheral equipment
- computer graphics
- terminal
- processor
- graphics scanner
- programmable logic controller
- input-output interface

Vocabulary Assistant

drawing	图纸	component	零件
conceptual	概念的	strength	强度
dynamic	动态的,动力学的	assembly	装配
interface	界面	geometry	几何学
discrete	分离的,离散的	stride	大步;进步
generative	生成的,生产的	incorporate	结合
digital	数字的	interpolation	插值,插入法
envelop	外壳	parameter	参数
software	软件	circuit	电路,回路
servo	伺服,伺服系统	driver	驱动器
memory	内存		

Answer the following questions:

1) What can CAD/CAM do for us?

2) In what field can we take advantages of CAD/CAM?

3) What is CAPP associated with?

4) What is CNC?

2. **The following are abbreviations of some terms associated with computer. Do you know what they mean? Find the full form in Column B and the Chinese meaning in Column C which matches the abbreviations in Column A. Then, can you add any?**

A	B	C
1. MCNC	a. Multimedia Terminal Adapter	① 编程系统信息协议
2. CAPP	b. Finite Element Method	② 无线接入点
3. MTA	c. Program System Information Protocol	③ 多媒体终端适配器
4. FEM	d. Microcomputer Numerical Control	④ 电源电压监视器
5. PSIP	e. System and External Memory Interface	⑤ 无线应用协议
6. SEMI	f. Wireless Application protocol	⑥ 计算机辅助工艺设计
7. WAP	g. Computer-Aided Process Planning	⑦ 有限元方法
8. SVS	h. Wireless Access Point	⑧ 接收信号处理器
9. RSP	i. Receive Signal Processor	⑨ 系统和外部存储器接口
10. WAP	j. Supply Voltage Supervisor	⑩ 微机型数控

1 — _____ — _____ ;   6 — _____ — _____ ;

2 — _____ — _____ ;   7 — _____ — _____ ;

3 — _____ — _____ ;   8 — _____ — _____ ;

4 — _____ — _____ ;   9 — _____ — _____ ;

5 — _____ — _____ ;   10 — _____ — _____ ;

Vocabulary Assistant

multimedia　多媒体　　　　adapter　适配器

protocol　协议　　　　　　external　外部的

multi-service　多服务　　　transport platform　传输平台

wireless　无线的　　　　　access　接入

signal　信号　　　　　　　supervisor　监视器

3. The following is a figure of the description of a PDM (Product Data Management) system consisting of CAD, CAM and CAPP. Can you tell the name of each part? And say something about the relationship of each part with a partner?

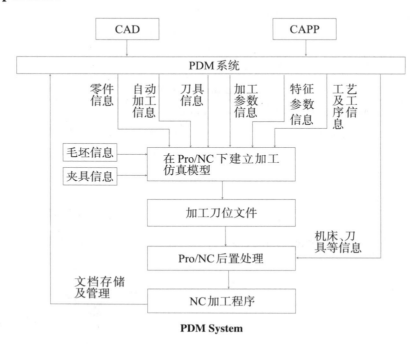

PDM System

Vocabulary Assistant

PDM (Product Data Management)　产品数据管理

part　零件

processing parameter　工艺参数

technique　技术

rough part　毛坯,半成品

emulation　仿真,模拟

file manipulation of tool position　加工刀位文件

processor　加工程序

management　管理,操纵

automatic processing　自动化加工

latent parameter　特征参数

working procedure　工序

jig　夹具

model　模型

storage of file　文档存储

machine tool　机床,工具机

4. The following are the features of CNC. Can you add any?

The flexibility of CNC has permitted the introduction of many convenient programming and operating features. Included among these features are the following:

★ editing of part programs at the machine

★ graphic display of the tool path to verify the tape

★ various types of interpolation: circular, parabolic, and cubic

★ use of specially written subroutines

★ manual data input (MDI)

★ local storage of more than one part program

5. The following is the list of the applications of CAD/CAM. Do you know what each of them refers to? Can you add any?

★ Programming for NC, CNC, and industrial robots

★ Design of dies and molds for casting

★ Design of tools and fixtures and EDM electrodes

★ Quality control and inspection

★ Process planning and scheduling

Vocabulary Assistant

graphic　图表的, 图画的

verify　查证, 核实; (在DOS命令下)打开/关闭在DOS操作期间的写文件校验开关

circular　传单, 通报; 循环的, 圆形的

parabolic　抛物线的　　　　　cubic　立方体的

subroutine　子程序　　　　　manual data input　手动数据输入

local storage　局部存储器　　electrode　电极

---

6. Match the definitions with the terms.

A. dimension	B. process	C. program	D. database

_____ 1) a usually large collection of data organized especially for rapid search and retrieval (as by a computer)

_____ 2) measurement of any sort (breadth, length, thickness, height, etc.)

_____ 3) series of actions or operations performed in order to do, make or
achieve something; perform operations on (something) in a computer

_____ 4) series of coded instructions to control the operations of a computer;
instruct a computer to do something by putting a program into it

Passage Reading

CAD / CAM / CAPP / CNC
计算机辅助设计/制造/工艺设计/计算机数控

CAD

1 Computer aided design (CAD) can be defined as using computers to aid the engineering design process by means of effectively creating, modifying, or documenting the part's geometrical modeling.

2 CAD is most commonly associated with the use of an interactive computer graphics system. The object of the engineering design is stored and represented in the form of geometric models. Geometric modeling is concerned with the use of a CAD system to develop a mathematical description of the geometry of an object. The mathematical description is called a model. There are three types of models (see the following figure): wire-frame models, surface models, and solid models, which are commonly used to represent a physical object. Wire-frame models, are the simplest method of modeling and are most commonly used to define computer models of parts. Surface models may be constructed using a large variety of surface features. Solid models are recorded in the computer result. Models in CAD can also be classified as being 2-D (two-dimensional) models, two-and-half-dimensional models, or 3-D (three-dimensional) models. Presently, many CAD systems can automatically generate the 2-D or 3-D FEM (finite-element method) meshes which are essential to FEM analysis.

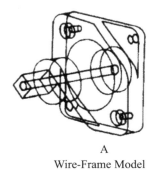
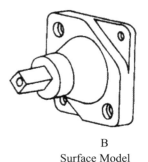
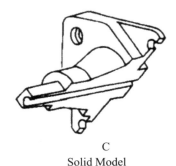

A	B	C
Wire-Frame Model	Surface Model	Solid Model

Three Types of Models

Questions: *What is CAD?*

Can you describe the function of CAD?

How many kinds of models does CAD have?

CAM

3 CAM can be defined as computer-aided manufacturing including decision-making, process and operational planning, software design techniques, and artificial intelligence. It also includes manufacturing with different types of automation (NC machine, NC machine centers, NC machining cells, NC flexible manufacturing systems), and different types of realization (CNC single unit technology, DNC group technology). The CAM covers group technology, manufacturing database, automation and tolerance.

4 When a design has frozen, manufacturing can begin. Computers have an important role to play in many aspects of production. Numerically controlled machine tools need a part program to define the components being made, and computer techniques exist to assist.

5 One of the most important manufacturing function is stock and production control. If the original design is done on a computer, obtaining lists of material requirements is straightforward. Standard computer data processing methods are employed to organize the work flow and order components when required.

Questions: *What is CAM?*

What is CAM associated with?

What is the function of CAM?

CAPP

6 CAPP can be defined as the functions which use computers to assist the work of process planners. The levels of assistance depend on the different strategies employed to implement the system. The highest level strategy is the highest goal of CAPP. That is, it will generate the process flow through computer rather than the process planners when the technical knowledge, expertise and working experience have been incorporated into computer programs. The database in a CAPP system based on the highest level strategy will be directly integrated with conjunctive systems, e.g. CAD and CAM. CAPP has been recognized as playing a key role in CIMS (Computer Integrated Manufacturing System).

7 Process planning is one of the basic tasks to be performed in manufacturing systems. The task of carrying out difficult and detailed process plans has traditionally been done by workers with a vast knowledge and understanding of the manufacturing process. Many of these skilled workers, who are called process planners, are either retired or close to retirement. However, there are no qualified young process planners who can take their place. An increasing shortage of process planners has been created. With the high pressure of serious competition in the world market, integrated production has been pursued as a way for companies to survive and succeed. Automated process planning systems have been recognized as playing a key role in CIMS. It is for reasons such as these that many companies look for computer aided process systems.

Questions: *What is CAPP?*

What is the relationship between CAPP and CAD/CAM?

What is the function of CAPP?

CNC

8 CNC stands for Computer Numerical Control and has been around since the early 1970s. Prior to this, it was called NC (numerical control). While people in most fields have never heard of this term, CNC has touched almost every form of manufacturing process in one way or another. People who work in manufacturing are very likely to deal with CNC on a regular basis.

9 Everything that an operator would be required to do with conventional machine tools is programmable with CNC machine. Once the machine is set up and made running, a CNC machine is quite simple to keep running. With some CNC machines, even the workpiece loading process has been automated.

10 The CNC control will interpret a CNC program and activate the series of commands in sequential order. As it reads the program, the CNC control will activate the appropriate machine functions, cause axis motion, and in general, follow the instructions given in the program. Along with interpreting the CNC program, the control has several other purposes. All current model CNC controls allow programs to be modified (edited) if mistakes are found. The CNC control allows special verification functions to confirm the correctness of the CNC program like tool length values. In general, the CNC control allows all functions of the machine to be manipulated.

Questions: *How does CNC develop?*

How does CNC program work?

Vocabulary Assistant

interactive 交互式的

wire-frame model 线框模型

solid model 实体模型

automatically 自动的

mesh 网络,网格结构

automation 自动化,自动控制

DNC 直接数字控制

straightforward 正直的;简单的,直截了当的

work flow 工作流程

geometric model 几何模型

surface model 表面模型

dimensional 维度的

FEM 有限元法

artificial intelligence 人工智能

FMS 柔性加工系统

database 数据库

order component 序分量

process flow　工艺流程　　　　　expertise　专业技术
conjunctive　相关的,相关联的　　CIMS　计算机集成制造系统
manufacturing process　制作过程　automate　使……自动化
activate　(使)起作用　　　　　　sequential　连续的,相继的
verification　证实,核实

1. Fill in the table below by giving the corresponding translation.

English	Chinese
computer monitoring and control	
	脱机应用
manufacturing support application	
	联机应用
integrated circuit computer aided design	
	电脑制图
computer-aided control system design	
	计算机设计自动化
computer-aided designing system.	
	人工智能

2. Match the items listed in the following two columns.

_____ 1) emerge　　　　　　　a. come out; come into view

_____ 2) software　　　　　　b. operating independently of an associated computer

_____ 3) numerical control　c. machine that processes things

_____ 4) computer interface　d. the programs used to direct the operation of a computer

_____ 5) off-line　　　　　　e. devices or programs designed to link one system to another

_____ 6) processor　　　　　f. control of a machine tool, or other devices in a manufacturing process by a computer

_____ 7) decline　　　　　　g. large in amount; considerable

_____ 8) substantial   h. become smaller, weaker, fewer, etc; diminish

_____ 9) graphics   i. to manage or utilize skillfully; to control or play upon by artful, unfair, or insidious means especially to one's own advantage

_____ 10) manipulate   j. the art or science of drawing a representation of an object on a two-dimensional surface according to mathematical rules of projection

3. Choose the best word and use the correct form from the box to complete the sentences.

| key   feed   concept   concern   database   terminal |

1) The moving belt _____ raw material into the machine.

2) The data can be distributed among _____ and computers.

3) It emerges as a _____ factor in CAD/CAM integration because it is the link between CAD and CAM.

4) Process planning is _____ with determining the sequence of individual manufacturing operations needed to produce a given part or product.

5) The main _____ of CAD/CAM systems is the generating of a common _____ which is used for all the design and manufacturing activities.

4. Pair Activities: Steven is a student majoring in Mechanical and Electrical Engineering. He is asked to explain CAD to a new student, Mike. (A: Mike, B: Steven)

A: Hi, Steven!

B: Hi, Mike!

A: Today, we have learned something about CAD, CAM and CAPP.

B: Oh, really? That sounds good. You know, those are very practical in work and everyday life.

A: So, I want to know more about CAD, for today our teacher introduced three types of models to us.

B: You mean wire-frame models, surface models, and solid models?

A: Yes, that's right! I like models very much. So, I want more about it. Could you please tell me?

B: Definitely! Look at the following figure. You see. It's three-dimensional model. As models for car, you can see the difference. Wire-frame model provides representation of a generalized part shape. It's the simplest one.

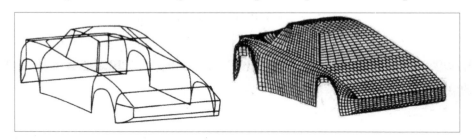

A. Wire-frame Model   B. Surface Model

Three-dimension Model

A: Wow, it is the simplest. Incredible!!

B: Sure. Surface models may be constructed using a large variety of surface features. And solid models are recorded in the computer result. Let's take the model of the car as an example, you can observe the picture, and then identify it.

A: It is so mysterious! But how can people make this?

B: Models in CAD also can be classified as being 2-D models, two-and-half-dimensional models, or 3-D models. After a particular design alternative has been developed, some form of engineering analysis must often be performed as a part of the design process.

A: Wonderful! I also learned that CAD and CAM are widely applied in mould design and mould making.

B: Right. As I told you, CAD allows you to draw a model on screen, then view it from every angle using 3-D animation and finally test it by introducing various parameters into the digital simulation models, such as pressure, temperature, impact, etc. So, CAM can allow you to control the manufacturing quality.

A: So it has a lot of advantages?

B: Right, such as shorter design times, lower cost, faster manufacturing, etc.

This new approach also allows to make last-minute changes to the mould for a particular part.

A: Then, do you know how it is used in designing the mould?

B: Ok, you know a CAD system consists of three basic components: hardware, software, users. The hardware components of a typical CAD system include a processor, a system display, a keyboard, a digitizer, and plotter. The software component of a CAD system consists of the programs which allow it to perform design and drafting functions.

A: You mean the user is the tool designer who uses the hardware and software to perform the design process?

B: Yes, based on the 3-D data of the product, the core and cavity have to be designed first. Usually the designer begins with a preliminary part design, which means the work around the core and cavity could change. Modern CAD systems can support this with calculating a split line for a defined draft direction, splitting the part in the core and cavity side and generating the run-off or shut-off surfaces. After the calculation of the optimal draft of the part, the position and direction of the cavity, slides and inserts have to be defined.

A: Then, what should be done?

B: Well, the positions and the geometry of the mould components—such as slides, ejection system, etc. —are roughly defined.

A: That's to say, with this information, the size and thickness of the plates can be defined and the corresponding standard mould can be chosen from the standard catalog.

B: Right, if there is no standard mould that will fit these needs, the standard mould which comes nearest to the requirements will be chosen and changed accordingly by adjusting the constraints and parameters so that any number of plates with any size can be used in the mould.

A: So, the use of the special 3-D mould design system can shorten development cycles and improve mould quality?

B: Great! In manufacturing, CAM is applied to reduce the cost and lead-time. The foundation for this type of manufacturing system is the use of CAD data to help in making key process decisions that ultimately improve machining precision and reduce non-productive time.

A: Ok, thank you, Steven! I've learned a lot. So I decide that I will read more about that and try to make one of my own one day.

B: I hope so. So, see you next time!

A: See you!

Answer the following questions:

1) What is three-dimensional model?

2) What are the three types of models mentioned in the dialogue?

3) How can people make those models?

4) Can CAD and CAM be applied in mould design and mould making? And how?

5) What are the advantages of CAD and CAM?

Vocabulary Assistant

practical 实际的   incredible 难以置信的

identify 辨别, 识别   3-D animation 3D 动画设计

digital simulation model 数字仿真模型

manufacturing quality 加工质量   low cost 低成本

consist of 包含, 包括   system display 系统显示器

keyboard 键盘   digitizer 数字化转换器

plotter 数据自动描绘器, 绘图机   core 核心, 铁心

preliminary 初步的   run-off 流出口, 流放口

shut-off 截流, 断流   optimal 最优的, 最佳的

ejection 排放   corresponding 相应的

constraint 限制   lead-time 交付周期, 研制周期

Further Reading

Flexible Manufacturing System
柔性制造系统

A flexible manufacturing system (FMS) is a system of numerical control (NC) machine tools, which has automated material handling and central computer control so that it has an automatic tool handling capacity. Because of its automatic tool handling capability and computer control, such a system can be continually reconfigured to manufacture a wide variety of parts. This is why it is called a flexible manufacturing system.

An FMS typically encompasses:

- Process equipment e.g. machine tools, assembly stations, and robots.
- Material handling equipment e.g. robots, conveyors, and AGVs (automated guided vehicles)
- A communication system
- A computer control system

Future FMS will contain many manufacturing cells, each cell consisting of robot serving several computer NC (CNC) machine tools or other stand-alone systems such as an inspection machine, a welder, an Electric Discharge Machining (EDM) machine, etc. The manufacturing cells will be located along a central transfer system, such as a conveyor, on which a variety of different workpieces and parts are moving. The production of each part will require processing through a different combination of manufacturing cells. In many cases, more than one cell can perform a given processing step. When a specific workpiece approaches the

FMS

required cell on the conveyor, the corresponding robot will pick it up and lead it onto a CNC machine in the cell. After processing in the cell, the robot will return the semi-finished or finished part to conveyor. A semi-finished part will move on the conveyor until it approaches a subsequent cell where its processing can be continued. The corresponding robot will pick it up and load it onto a machine tool. This sequence will be repeated along the conveyor, until, at the route, there will be only finished parts moving. Then they could be routed to an automatic inspection station and subsequently unloaded from the FMS. The coordination among the manufacturing cells and the control of the part's flow on the conveyor will be accomplished under the supervision of the central computer.

The advantages of FMS

- Increased productivity
- Shorter preparation time for new products
- Reduction of inventory parts in the plant
- Saving of labor cost
- Improved product quality
- Attracting skilled people to manufacturing (since factory work is regarded as boring and dirty)
- Improved operator's safety

Additional economic savings may be from such things as the operator's personal tools, gloves, etc. Other savings are in locker rooms, showers, and cafeteria facilities — all representing valuable plant space, which will not require enlarging if company growth is achieved with flexible automation systems.

The advantages of CNC

CNC machines do not necessarily produce components more accurately than NC machines, and are only faster in operation owing to the speed with which the program blocks are transmitted to the control unit. CNC reduces the amount of non-chip-producing time by several ways: selecting speeds and feeds, making rapid moves between surfaces to be cut, using automatic fixtures, changing automatic tool, controlling the coolant, in-process gagging, loading and unloading the part. These factors, plus the fact that it is no longer necessary to train machine operators, have resulted in considerable savings throughout the entire manufacturing process and caused tremendous growth in the use of CNC. Some of the major advantages of

CNC are as follows:

★ The amount of input information is reduced.

★ There is automatic or semiautomatic operation of machine tools. The degree of automation can be selected as required.

★ Flexible manufacturing of parts is much easier. Only the part programming needs changing to produce another part.

★ Storage space is reduced. Simple work holding fixtures are generally used, reducing the number of jigs or fixtures which must be stored.

★ Small part lots can be run economically. Often a single part can be produced more quickly and better by CNC.

★ Nonproductive time is reduced. More time is spent on machining the part, and less time is spent on moving and waiting.

Over to you:

1) What is FMS?

2) What will future FMS contain? How will it work?

3) What are the advantages of FMS?

4) What are the advantages of CNC?

Vocabulary Assistant

reconfigure 改造,更换部件	encompass 包含,包括
conveyor 输送机	welder 焊机
coordination 协作,协调	program block 程序块
fixture 夹具	coolant 冷却剂
gagging 冷矫正,成型	part lots 零件组,零件群

Unit 5

Computer Integrated Manufacturing System
计算机集成制造系统

Introduction

COIS

Computer Integrated Manufacturing, known as CIM, is the phrase used to describe the complete automation of a manufacturing plant, with all processes functioning under computer control and digital information tying them together. It includes CAD/CAM (computer-aided design/computer-aided manufacturing), CAPP (computer-aided process planning), CNC (computer numerical control), DNC (direct numerical control machine tools), FMS (flexible machining systems), AS/RS (automated storage and retrieval systems), AGV (automated guided vehicles), use of robotics and automated conveyance, computerized scheduling and production control, and a business system integrated by a common database.

CIM is the integration of all the processes necessary to manufacture a product through computer technology. In manufacturing, CIM in its fullest implementation can integrate all manufacturing information, not only the manufacturing data for products, but also the business procedures, corporate goals, and management structure of a manufacturing enterprise. A simulation based on control system is utilized for computer integrated control of the work cells.

CIMS can not only improve the design of product innovation and enlarge the technological content of the product, but also establish a computer integrated manufacturing environment for an enterprise from product design, manufacturing, to management. With the support of CIMS, a comprehensive feedback in the market can be fed back to the subsystem of automatic design (CAD/CAPP /CAM/). Then, with the help of PDM and CAD systems, the designers of products can make use of past experiences in designing and promoting the designs to meet the market demand.

CIMS generally includes six components:

Management information system (MIS) deals with the production planning and control, management, sales management, procurement management, financial management, and other functions dealing with the information of the production.

Subsystem of engineering design application consists of CAD, CAPP and the process of NC programming to help the design, preparation and handling of the products.

Workshop management and the subsystem of automation application is also known as CAM subsystem, which includes a variety of manufacturing equipment of different automation and subsystems used to support the manufacturing functions of enterprises.

Computer-aided quality assurance system (CAQ) deals with the formulation of quality management plan, carries out quality management and handles the information about quality.

Management subsystem of data inventory deals with the data of CIMS to realize the integration and sharing of data.

Computer and network subsystem transfers the information between the various subsystems and internal information within subsystems and realizes the CIMS data transmission and communication systems.

1. The following are the potential benefits of CIM. Do you know the meaning of each? Can you add any?

◆ Shorter time to market new product

◆ Increase in manufacturing productivity

◆ Shorter customer lead time

◆ Shorter vendor lead time

◆ Improved quality

◆ Improved customer service

◆ Reduced inventory level

◆ Greater flexibility and responsiveness

◆ Lower total cost

Vocabulary Assistant

computer integrated manufacturing　计算机集成制造
numerical　数字的　　　　　　　storage　储存
conveyance　输送,传导　　　　　enterprise　企业
simulation　仿真,模拟　　　　　utilize　利用
innovation　创新,革新　　　　　feedback　回应,反馈
sales management　销售管理　　procurement management　采购管理
financial　财政的,金融的　　　formulation　明确表达
transmission　传送

Answer the following questions:

1) What is CIM?

2) When CIMS is applied in a factory, will its manufacturing efficiency be increased? And how? In what aspects?

3) Can you give successful examples of CIMS application?

2. Look at the following expressions. Do you know their meanings? Can you add any?

★ Computer Integrated Manufacturing (CIM)

★ Computer Numerical Control (CNC)

★ Computer-aided design (CAD)

★ Computer-aided manufacturing (CAM)

★ Automated storage and retrieval systems (AS/RS)

★ Just-in-time (JIT)

3. The following is a chart of CIMS. Do you know what each activity refers to? Then work with a partner to discuss how each subsystem works in the integration of a corporation's manufacturing activities.

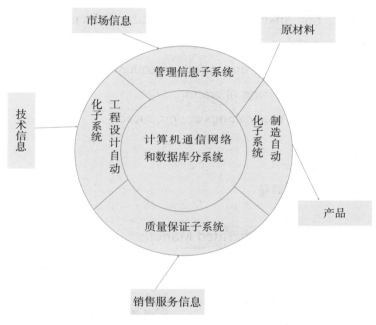

The Chart of CIMS

Vocabulary Assistant

computer telecommunication network system　计算机通信网络系统

manufacturing automation　制造自动化

engineering design　工程设计　　　　guarantee　保证

distribution　销售　　　　　　　　　raw material　原材料

4. Match the definitions with the terms.

A. CIM	B. modern manufacturing
C. manufacturing plant	D. numerical control

_____ 1) The phrase used to describe the complete automation of a manufacturing plant, with all processes functioning under computer control and digital information tying them together

_____ 2) The activities encompassing all of the activities and processes necessary to convert raw materials into finished products, deliver them to the market, and support them in the field

_____ 3) An arrangement of machines or sequence of operations involved in a single manufacturing operation

_____ 4) Factory where many things are produced

Passage Reading

Computer Integrated Manufacturing System
计算机集成制造系统

1 Computer integrated manufacturing (CIM) is the term used to describe the most modern approach to manufacturing. Although CIM encompasses many of the other advanced manufacturing technologies such as computer numerical control (CNC), computer-aided design/computer-aided manufacturing (CAD/CAM), robotics, and just-in-time delivery (JIT), it is more than a new technology or a new concept. Computer-integrated manufacturing is actually

an entirely new approach to manufacturing a new way of doing business.

> **Questions:** *What can CIM do as far as you know?*
> *What can CIM be used for?*

2 To understand CIM, it is necessary to begin with a comparison of modern and traditional manufacturing. Modern manufacturing encompasses all of the activities and processes necessary to convert raw materials into finished products, deliver them to the market, and support them in the field. These activities include the following:

(a) identifying a need for a product.

(b) designing a product to meet the needs.

(c) obtaining the raw materials needed to produce the product.

(d) applying appropriate processes to transform the raw materials into finished products.

(e) transporting product to the market.

(f) maintaining the product to ensure a proper performance in the field.

> **Question:** *What does modern manufacturing encompass?*

3 This broad, modern view of manufacturing can be compared with the more limited traditional view that focuses almost entirely on the conversion processes. The old approach separates such critical pre-conversion elements as market analysis research, development, and design for manufacturing, as well as such after-conversion elements as product delivery and product maintenance. In other words, in the old approach to manufacturing, only those processes that take place on the shop floor are considered manufacturing. This traditional approach of separating the overall concept into numerous stand-alone specialized elements was not fundamentally changed with the advent of automation. While the separate elements themselves became automated (i. e. computer-aided drafting and design (CADD) in design and CNC in machining), they remained separate. Automation alone did not result in the integration of these islands of automation.

Questions: *What is the traditional approach to manufacturing work?*
What are the differences between traditional manufacturing approach and modern one?

4 With CIM not only are the various elements automated, but the islands of automation are all linked together or integrated. Integration means that a system can provide complete and instantaneous sharing of information. In modern manufacturing, integration is accomplished by computers. With this background, CIM can now be defined as the total integration of all manufacturing elements through the use of computers.

Vocabulary Assistant

robotics   机器人技术   delivery   交付
comparison   比较   pre-conversion   转变前
after-conversion   转变后   advent   出现,到来
instantaneous   瞬间的,即时的

---

1. Fill in the table below by giving the corresponding translation.

English	Chinese
advent	
	机器人技术
conversion	
	产品维护
critical	
	原材料
maintain	
	成品
accomplish	
	即时交付

2. Find the definition in Column B which matches the words in Column A.

A	B
_____ 1) transform	a. with various parts fitting well together
_____ 2) analysis	b. new and not yet generally accepted
_____ 3) integrated	c. study of sth. by examining its parts and their relationship
_____ 4) obtain	d. in the natural state
_____ 5) advanced	e. completely change the appearance of sth.
_____ 6) concept	f. idea; general notion
_____ 7) raw	g. get sth.; come to own or possess sth.
_____ 8) manufacture	h. make goods on a large scale using machinery
_____ 9) automation	i. use of automatic equipment and machines to do work previously done by people
_____ 10) encompass	j. include or compromise sth.; surround

3. Use the correct form of the words from the box to complete the sentences.

advent	deliver	achieve	approach	convert

1) Modern manufacturing encompasses all of the activities to _____ raw materials into finished product.
2) With the _____ of the computer age, manufacturing has developed full circle and began as a totally integrated concept.
3) Computer-integrated manufacturing is actually an entirely new _____ to manufacturing a new way of doing business.
4) The products _____ to the market were manufactured by the workers.
5) In modern manufacturing, integration is _____ by computers.

4. Pair Activities: Professor White is asked to explain CIM to the new students majoring in Mechanical and Electrical Engineering. (A: a new student, B: Professor White)

A: Professor, what is CIM?

B: CIM can now be defined as the total integration of all manufacturing elements through the use of computers. It is the process of automating various functions in a manufacturing company (business, engineering, and production) by integrating the work through computer networks and common database.

A: Is it useful in modern manufacturing firm?

B: Sure. CIM is a critical element in the competitive strategy of global manufacturing firms because it can lower costs and improve delivery times and quality.

A: When was this term developed?

B: It was developed in 1974 by Joseph Harrington as the title of a book he wrote about tying islands of automation together through the use of computers.

A: Then it has taken many years for CIM to develop as a concept.

B: Yes, but integrated manufacturing is not really new. In fact, integration is where manufacturing actually began.

A: How many stages has manufacturing had?

B: Four stages.

A: What are they?

B: They are manual manufacturing, mechanization or specialization, automation and integration.

A: Then, could you explain the major components of CIM?

B: Sure, just as I've told you at the very beginning, CIM system shows how the various machines and processes used in the conversion process are integrated.

A: What processes?

B: For example, product design, process planning, programming automated machines, production planning, production, shipping, receiving, storing, finance, marketing are all integrated in a CIM system.

A: Yes, we've got it. Thank you so much, Professor!

B: My pleasure.

> **Answer the following questions:**
> 1) *How did the term CIM develop?*
> 2) *How many stages has manufacturing had?*
> 3) *What benefits can a fully integrated manufacturing firm get?*

5. Application

CIM can be applied in many areas. In the area of engineering design automation, CIM can be used to develop the product and increase its productivity. It is especially useful in developing the structurally-complicated product and product of advanced technology. CIM can ensure the design quality of product, shorten the cycle of product design and craftwork design, and, therefore, can update the product and satisfy the requirement of customers. In the area of manufacturing automation, CIM can strengthen the flexibility of product, shorten the cycle of product manufacturing, and increase the productivity. In the field of management administration, CIM can make the management policy and management of product tend to be scientific. It can also promote the enterprise to give quoted price in market competition quickly and accurately.

The following figure is the major components of CIM according to the above dialogue. Do you know what each of them stands for? How are the different activities integrated? What are the technological advantages of CIMS? Then work with a partner to discuss how CIMS is used in modern manufacturing plant.

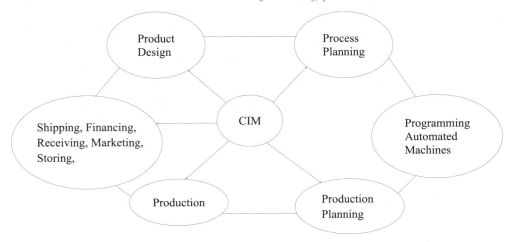

Major Components of CIM

Over to you:

True or False Questions

_____ 1) The traditional manufacturing have already encompassed all of the activities and processes necessary to convert raw materials into finished products, deliver them to the market, and support them in the field.

_____ 2) The modern manufacturing is more advanced than the traditional one because it focuses on the conversion processes to a large degree.

_____ 3) The processing activities include identifying a need for a product; designing a product to meet the needs; obtaining the raw materials needed to produce the product; applying appropriate processes to transform the raw materials into finished products; transporting product to the market; maintaining the product to ensure a proper performance in the field.

_____ 4) In the traditional manufacturing, only those processes that happened on the shop floor were considered manufacturing.

Vocabulary Assistant

structurally-complicated product 结构复杂的产品

product of advanced technology 高新技术产品

cycle 周期,循环   craftwork design 工艺设计

update 更新   management administration 经营管理

management policy 经营方针,管理策略

quoted price 报价   market competition 市场竞争

Further Reading

Problems Associated with CIM
与计算机集成制造相关的问题

As with any new philosophy that requires major changes to the status quo, CIM is not without problems. The problems associated with CIM fall into three major categories: technical problems, cultural problems, and business-related problems.

These types of problems have hindered the development of CIM over the years and will have to be overcome for CIM to achieve widespread implementation.

(1) Technical problems of CIM

As each island of automation began to evolve, specialized hardware and software for that island were developed by a variety of producers. This led to the same type of problem that has been experienced in the automotive industry. One problem in maintaining and repairing automobiles has always been the incompatibility of spare parts among various makes and models. Incompatibility summarizes in a word the principal technical problem inhibiting the development of CIM. Consider the following example. Supplier A produces hardware and software for automating the design process. Supplier B produces hardware and software for automating such manufacturing processes as machining, assembly, packaging, and materials handling. Supplier C produces hardware and software for automating processes associated with market research. This means a manufacturing firm may have three automated components, but on systems produced by three different suppliers. Consequently, the three systems are not compatible. They are not able to communicate among themselves. Therefore, there can be no integration of the design, production, and market research processes.

An effort known as manufacturing automation protocol (MAP) is beginning to solve the incompatibility of hardware and software produced by different suppliers. As MAP continues to develop, the incompatibility problem will finally be solved and full integration will be possible among all elements of a manufacturing plant.

(2) Cultural problems of CIM

Computer-integrated manufacturing is not just new manufacturing technology; it is a whole new approach to manufacturing, a new way of doing business. As a result, it includes significant changes for people who were educated and are experienced in the old ways. As a result, many people reject the new approach represented by CIM for a variety of reasons. Some simply fear the change that it will bring in their working lives. Others feel it will make them lose their job. In any case, the cultural problems associated with CIM will be more difficult to solve than the technical problems.

(3) Business-related problems of CIM

Closely tied to the cultural problems are the business problems associated with CIM. The accounting problem comes first. Traditional accounting practices do not work with CIM. There is no way to prove CIM based on traditional accounting practices. Traditional accounting practices base cost-effectiveness studies on direct labor savings whenever a new approach or new technology is "made". However, the savings that result from CIM are more closely tied to indirect factors, which are more difficult to count. Therefore, it can be difficult to convince traditional business people who are used to relying on traditional accounting practices to see that CIM is an approach worth the investment.

Vocabulary Assistant

status quo   现状   fall into   分成
implementation   执行,实现
automotive industry   汽车工业,汽车制造业
incompatibility   不相容性   summarize   概述
inhibit   抑制,禁止   packaging   包装
materials handling   材料处理   compatible   相兼容的,能共处的
cost-effectiveness   成本效益   convince   说服

Over to you:

1) What are the problems associated with CIM?

2) In technical aspect, what is the most difficult one?

3) What's the cultural problem of CIM?

Unit 6

Integrated Circuits
集成电路

Introduction

Integrated Circuits are usually called ICs or chips. They are complex circuits which have been etched onto tiny chips of semiconductor (silicon). The chip is packaged in a plastic holder with pins spaced on a certain type of grid which will fit the holes on strip board and breadboards. Very fine wires inside the package link the chip to the pins.

The integrated circuit is a very advanced electric circuit. It is made from different electrical components such as transistors, resistors, capacitors and diodes, which are connected to each other in different ways. These components have different behaviors.

transistor

The transistor acts like a switch. It can turn electricity on or off, or it can amplify current. It is used, for example, in computers to store information, or in stereo amplifiers to make the sound signal stronger.

The **resistor** limits the flow of electricity and gives us the possibility to control the amount of current that is allowed to pass. Resistors are used to control the volume in television sets or radios.

resistor

capacitor

The **capacitor** collects electricity and releases it all in one quick burst; such as in cameras where a tiny battery can provide enough energy to fire the flashbulb.

The **diode** stops electricity under some conditions and allows it to pass only when these conditions change. This is used in, for example, photocells where a light beam that is broken triggers the diode to stop electricity from flowing through it.

The diode is a two-terminal device. Diodes have two active electrodes, and most are used for their unidirectional current property. Diodes allow electricity to flow in only one direction.

diode

1. The following are the terms about integrated circuits. Do you know what each of them refers to? Can you add any?

circuit element	transistor	diode
wafer	resistor	breadboard
capacitor	functional block	monolithic

Vocabulary Assistant

integrated circuit 集成电路	etch 蚀刻
chip 芯片	semiconductor 半导体

機電英語(第二版)

silicon　硅	pin　栓,管(脚)
grid　格子,网格,点阵	breadboard　电路试验板
transistor　晶体管	resistor　电阻器
capacitor　电容器	diode　二极管
switch　开关	amplify　扩大,放大
stereo amplifier　立体声扩音器	release　释放
burst　爆裂,爆发	battery　电池
flashbulb　[摄]闪光灯泡,镁光灯	photocell　光电池,光电管
light beam　线偏振光束	trigger　引发,触发
wafer　晶片	functional block　功能块
monolithic　单块集成电路	

Answer the following questions:

1) What does an IC consist of?

2) What is transistor?

3) What is resistor?

4) What is the function of capacitor?

5) What is the function of diode?

2. **The following are the pictures of the components in an integrated circuit. Match the term with the corresponding picture.**

1 _____

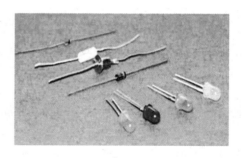

2 _____

86

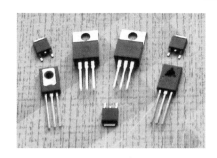

3 _____   4 _____

A. transistor   **B. resistor**   **C. capacitor**   **D. diode**

3. The following (Fig. 6-1) is the classifications of integrated circuits. Work with a partner to discuss how it is classified and what each category includes.

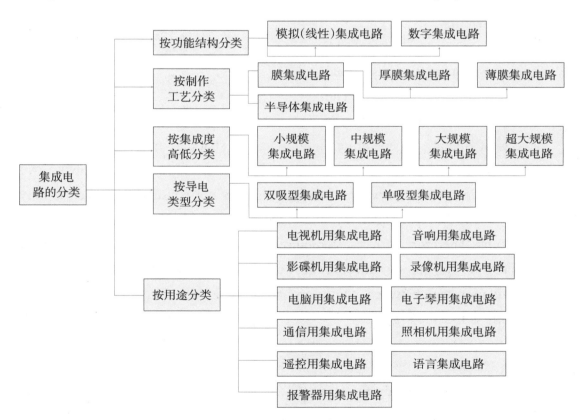

Fig. 6-1 集成电路的分类

Vocabulary Assistant

linear IC   模(线性)集成电路   technics   工艺
film   膜体   conductive   导电
double-suction type   双吸型   single-suction type   单吸型
audio   音响   player   影碟机
electronic organ   电子琴   communication   通信
remote control   遥控   alarm   报警器

4. The following are four basic ICs. Read the description of each term. Then discuss with a partner to identify the differences among them.

Digital Circuits

Digital Circuits deal with the dispersive, non-continuous signal (known as digital signal). The study of digital circuits concerns

Fig. 6-2 Digital Circuit Waveform

the digital signal's generation, amplification, shaping, transmission, control, memory and counting, and so on. Digital circuits mainly have the following two characteristics: First, the work of digital signal is not continuous, which is only about the existence of signal and the high-level or low-level of the electricity in the circuits. Therefore, it has less requirements of accuracy in the circuits, and it is suitable for integration. Second, the object of the study of digital circuits is the logic relationship between the input and output of the circuit. The main waveform is like the Fig. 6-2.

Analog Circuits

Analog circuits study the size of the current in a certain period of time. Its process is a continuous

Fig. 6-3 Analog Circuit Waveform

physics. It mainly applies to the field of the driving terminal load of signal amplification disposal. The major method is to set working points. Analog circuits can handle all kinds of signals, such as magnifying the signal from the spacecraft tens of thousands of times. The waveform is like Fig. 6-3.

Differential Circuit

Differential circuits (Fig. 6-4) can convert a rectangular wave into pulse wave. The output waveform of the circuit only re-

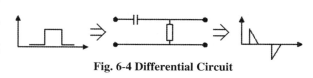

Fig. 6-4 Differential Circuit

flects the mutational part of the input waveform. That is, it is only when the muta-tion of the input waveform occurs that output will occur. The width of the shape of the output pulse wave is related to R * C (the circuit time constant). The small-er the R * C is, the sharper the pulse waveform becomes; on the contrary, it be-comes wider. The R * C of the circuit must be far less than the width of the input waveform, or the waveform transformation will lose its function and become the general R* C coupling circuit. However, generally speaking, it would be all right if the R * C is less than or as much as 1 / 10 of the width of the input waveform.

Integral Circuit

Integral circuits (Fig. 6-5) can convert rectangular pulse wave into sawtooth or triangular wave. It can

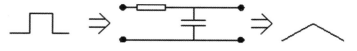

Fig. 6-5 Integral Circuit

also convert sawtooth wave into parabolic wave. The working principle of the cir-cuit is very simple for it is all based on the capacitor discharge principle. Here it mainly refers to R * C — the circuit time constant. The condition that constitutes integral circuits is that the circuit time constant must be 10 times more than or as much as the width of the input waveform.

Vocabulary Assistant

digital circuit　数字电路　　　　　dispersive　分散的,散布的

amplification　扩大,增幅　　　　　counting　计算

integration　集合;集成　　　　　　object　目标

logic 逻辑(学)	input 输入
output 输出	waveform 波形
analog circuit 模拟电路	physics 物理过程,物理现象
magnify 扩大,夸大	differential circuit 微分电路
mutational 变化的,转变的,突变的	time constant 时间常数,时间恒量
integral circuit 积分电路	sawtooth 锯齿
triangular 三角形	parabolic wave 抛物线形波

5. Match the definitions with the terms.

A. resistor   B. transistor   C. diode   D. microprocessor

_____ 1) an electronic device that has two electrodes or terminals and is used especially as a rectifier

_____ 2) a computer processor contained on an integrated-circuit chip

_____ 3) a device that has electrical resistance and that is used in an electric circuit for protection, operation, or current control

_____ 4) a solid-state electronic device that is used to control the flow of electricity in electronic equipment and usually consists of a small block of a semiconductor with at least three electrodes

Passage Reading

Integrated Circuits
集成电路

1 Our world is full of integrated circuits (ICs). You find several of them in computers. For example, most people have probably heard about the microprocessor. The microprocessor is an integrated circuit that processes all information in the computer. It keeps track of what keys are pressed and if the mouse has been moved. It counts numbers and runs programs, games and the operating system. Integrated circuits are also found in almost every modern electrical device such as cars, television sets, CD players, cellular phones. But what is an integrated circuit?

2 IC is a combination of a few interconnected circuit elements such as transistors, diodes, capacitors, and resistors. These elements are produced in a single manufacturing process on the bearing structure and they are intended to perform a definite function involved in converting information.

> **Questions:** *What is microprocessor?*
>
> *What is IC?*

3 The inseparably associated and electrically interconnected components that make up an IC are called integrated elements. If an integrated circuit includes only one type of components, such as only diode or resistors, it is said to be an assembly or set of components.

4 The principle of integrated circuit elements lies in the following. A great number of "sets" are produced simultaneously on a wafer. Each set contains all the components such as transistors, diodes, and resistors that are interconnected with short fine metallic stripes deposited on the wafer surface. They make up an appropriate functional block. Each IC component is a ready integrated circuit. All ICs are regularly distributed on the wafer surface.

> **Questions:** *What are integrated elements?*
>
> *What is the principle of integrated circuit elements?*
>
> *How do ICs distribute on the wafer surface?*

5 The manufacturing techniques used for ICs can be divided into two main types: film technique and monolithic technique. And ICs can be classified by function into two: circuits to be applied in digital systems and those to be applied in linear systems. The digital ICs are employed mostly in comput-

ers, electronic counters, frequency synthesizers and digital instruments. And the analog, or linear ICs operate over a continuous range, and include such devices as operational amplifiers.

Questions: *How to classify the ICs by function?*
How are the digital ICs employed?

6 Many ICs are static sensitive and can be damaged when you touch them because your body may have become charged with static electricity, from your clothes for example. Static sensitive ICs will be supplied in antistatic packaging with a warning label and they should be left in this packaging until you are ready to use them.

7 The invention of IC is a great revolution in the electronic industry. Sharp size, weight reductions are possible with these techniques; and more importantly, high reliability, excellent functional performance, low cost and low power dissipation can be achieved. ICs are widely used in the electronic industry.

Questions: *Why are many ICs static sensitive?*
What are the achievements of the IC's invention?

Vocabulary Assistant

track　轨迹	cellular phone　便携式电话
interconnect　使互相连接	simultaneously　同时地
stripe　条纹	deposit　堆积,存放
monolithic technique　单块技术	linear　线性的
counter　计算器	frequency synthesizer　频率合成器
analog　类似物,相似体	static　静力的,静态的
sensitive　敏感的,灵敏的	static electricity　静电

antistatic 抗静电的	label 标签
high reliability 高可靠性	dissipation 消散，分散

1. Fill in the table below by giving the corresponding translation.

English	Chinese
dissipation	
	集成电路
simultaneously	
	高可靠性
static electricity	
	电容器
stereo amplifier	
	计算器
frequency synthesize	
	单块技术

2. Find the definition in Column B which matches the words in Column A.

A | B

_____ 1) combination a. clear; not doubtful

_____ 2) manufacture b. arrange or order by classes or categories

_____ 3) integrate c. the purpose that something is made for

_____ 4) definite d. that of uniting to form one.

_____ 5) convert e. any of the parts of which sth. is made

_____ 6) component f. complete or drastic change of method, conditions, etc.

_____ 7) classify g. the process of producing

_____ 8) revolution h. a piece of equipment intended for a particular purpose

_____ 9) function i. to unite with something else

_____ 10) device j. change (sth.) from one form or use to another

3. Use the correct form of the words from the box to complete the sentences.

> microprocessor   resistor   electric   connect   technology   invent

1) Integrated circuits are found in almost every modern _____ device such as cars, television sets, CD players, cellular phones.

2) A _____ restricts the flow of current, for example, to limit the current passing through an LED (Light Emitting Diode).

3) The _____ of the transistor was considered a revolution in 1947.

4) When building a circuit, it is very important that all _____ are intact. If not, the electrical current will be stopped on its way through the circuit, making the circuit fail.

5) Computer chip _____ is in all sorts of everyday items, from traffic lights to computers.

6) The most sophisticated chip is a _____, which is the most complex manufactured product on earth.

4. Pair Activities: Steven is a new student majoring in Mechanical and Electrical Engineering. He is asking Professor Smith some questions about integrated circuit. (A: Steven B: Professor Smith)

A: Good morning, Professor Smith. Could I ask you some questions?

B: Ok, go ahead.

A: Today, we have learned integrated circuits, and I want to know more about it. For example, could you please tell me the history of integrated circuits?

B: OK. The first integrated circuits were created in the late 1950s. In 1958 Jack Kilby had his first integrated circuit. It was tested and it worked perfectly!

The integrated circuit's mass production capability, reliability, and building-block approach to circuit design ensured the rapid adoption of standardized ICs in place of designs using discrete transistors.

A: I learned that there are four basic ICs. But I have some questions about digital ICs and analog ICs. Could you please explain them?

B: Digital circuits are circuits dealing with signals restricted to the extreme limits of zero and some full amount. This stands in contrast to analog circuits, in which signals are free to vary continuously between the limits imposed by power supply voltage and circuit resistances.

A: Oh, I see.

B: What's more, the impact of integrated circuits on our lives has been enormous. It has nearly been used in all fields, computer industry, electrical productions, IC cards, motor industry, communication, industrial control system and so on.

A: Then, what makes it so popular?

B: Well, IC microcomputers are smaller and more versatile than previous control mechanisms. They allow the equipment to respond to a wider range of input

and produce a wider range of output. They can also be reprogrammed without having to redesign the control circuitry. Integrated circuit microcomputers are so inexpensive that they are even found in children's electronic toys.

A: Really! Sounds incredible! Thank you.

B: You are welcome.

Vocabulary Assistant

LED (Light Emitting Diode)　发光二极管
in place of　替代　　　　　　　　　circuitry　电路,电路学

> **Answer the following questions:**
> 1) What are digital ICs and analog ICs?
> 2) Look at the following chart, and discuss how integrated circuits is applied in our daily life.

其他 2.2%
通信 20.0%
消费电子 26.1%
IC卡 0.9%
工控 7.1%
汽车电子 1.3%
计算机 42.4%

The Application of Integrated Circuits in Different Field

5. Application

Do you know in what fields integrated circuit can be applied? Discuss with a partner to explain how integrated circuit is used in the following fields.

A   B

C   D

Over to you:

True or False Questions

_____ 1) Integrated circuits perform a definite function involved in transforming information.

_____ 2) If an integrated circuit includes only one type of components, it is said to be an assembly.

_____ 3) Each IC of components is a ready integrated circuit and all ICs are irregularly distributed on the wafer surface.

_____ 4) Integrated elements can be applied in digital systems and linear systems.

_____ 5) The invention of IC is very important to the electronic industry.

Further Reading

Circuits
电　路

An electrical circuit is a network that has a closed loop, giving a return path for the current. A network is a connection of two or more components, and may not necessarily be a circuit. A network that also contains active electronic components is known as an electronic circuit. Such networks are generally nonlinear and require more complex design and analysis tools.

An electric circuit often consists of four parts: the source or power supply such as a battery, the conductor or wires, the control device such as a switch, and the

load. The load is a device or a machine. Within the load the actual energy conversion takes place. The lamp, the motor are common examples of electric loads.

A lamp connected across a dry cell is an example of a simple electrical circuit. Current flows from the negative terminal of the cell, through a lamp, to the positive terminal. The action of the cell can provide a "regenerative" path for the flow of electrons to be maintained through the negative terminal once again.

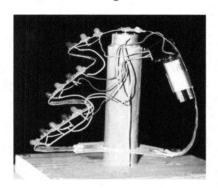

As long as this electrical pathway remains unbroken at any point, it is closed circuit and current flows; but if the pathway is ever broken, it becomes at once an open circuit and no current can flow.

Series circuits and parallel circuits are two main types of circuits connection. When electrical devices are connected so that the current is not divided at any point, they are said to be connected in series. The current in every part of this kind of circuit is the same. In ordinary house lighting, for instance, lamps are connected in parallel, each lamp filament representing an independent path from the minus main wire to the plus wire. In parallel circuits the total current is equal to the sum of the current that are passing through the branches of that of another.

Many practical circuits are arranged in series-parallel. Such circuits make it possible to combine the different voltage characteristic of a series circuit with the different current characteristic of a parallel circuit within a single network. The condition is particularly advantageous when it is necessary to operate loads that have different voltage and current requirements from the same source of energy.

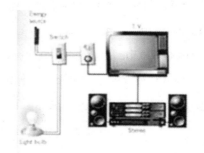

To design any electrical circuit, it is necessary to predict the voltages and currents at all places within the circuit. Linear circuits, that is, circuits with the same input and output frequency, can be analyzed by hand using complex number theory.

Other circuits can only be analyzed with specialized software programs or estimation techniques.

Circuit simulation software, such as VHDL and HSPICE, allows engineers to design circuits without worrying about time, cost and risk of error involved in building circuit prototypes.

The following are some of practical problems concerning circuits. Work with a partner to understand the meaning of them.

Dangers of Electrical Energy

Only specialized electricians are allowed to work on electric installations or equipment in accordance with electrotechnical regulations. Electrical shocks may cause injuries and burns.

Causes

- Contact, touching or immediate vicinity of non-insulated or defectively insulated parts that are energized.
- Exposed, electrically conductive parts after failure of insulation.
- Defective performance and insufficient safety check after maintenance work.
- Use of wrong fuses.
- Short-circuit.

Protective Measures

- Do not touch live parts.
- Before carrying out tests or maintenance work, Make sure any machine and system parts to be checked, maintained or repaired.
- Check de-energized parts for voltage first. Then earth and short-circuit them and insulate any adjacent live parts.
- Regularly check electric equipment.
- Wear safety clothing (plastic or rubber gloves, shoes with thick crepe soles).
- Only use insulated tools.
- Replace blown fuses by fuses of the same type only.

Vocabulary Assistant

loop　环,圈,弯曲部分

conductor　导线,导体

conversion　转换

dry cell　干电池

negative terminal　负极接线柱

positive　正的

regenerative　再生的

closed circuit　闭合电路

open circuit　断路,开路

series circuit　串联电路

parallel circuit　并联电路

filament　细丝,灯丝

minus　负的

plus　正的

series-parallel　串并联

estimation　判断

installation　安装

accordance　相符,一致

electrotechnical　电气工艺学的

vicinity　附近(电工技术的)

defectively　有缺陷地,缺乏地

energize　给予……电压,使电流入

expose　暴露

fuse　保险丝;导火线

short-circuit　短路

de-energize　断路

earth　把(电线或其他导体)接地

adjacent　邻近的

crepe sole　绉胶鞋底

blown fuse　熔断的保险丝

Over to you:

1. How many parts are there in the circuits? What are they?

2. What is closed circuit?

3. What is open circuit?

4. What are the two main types of circuits?

5. Why are many practical circuits arranged in series-parallel?

Unit 7

The Electrical Discharge Machining Process
电火花加工

Introduction

Electrical Discharge Machining (EDM) is a machining method primarily used for hard metals or those that would be impossible to machine with traditional techniques. One critical limitation, however, is that EDM only works with materials that are electrically conductive. EDM is especially well-suited for cutting intricate contours or delicate cavities that would be difficult to produce with a grinder, an end mill or other cutting tools. Metals that can be machined with EDM include alloy, hardened tool-steel, titanium, carbide nickel.

EDM is sometimes called "spark machining" because it removes metal by producing a rapid series of repetitive electrical discharges. These electrical discharges are passed between an electrode and the piece of metal being machined. The small amount of material that is removed from the workpiece is flushed away with a continuously flowing fluid. The repetitive discharges create a set of successively deeper craters in the workpiece until the final shape is produced.

There are two primary EDM methods: ram EDM and wire EDM. The primary difference between the two involves the electrode that is used to perform the machining. In a typical ram EDM application, a graphite electrode is machined with traditional tools. The now specially-shaped electrode is connected to the power source, attached to a ram, and slowly fed into the workpiece. The entire machining operation is usually performed while submerged in a fluid bath. The fluid serves the following three purposes:

★ flushes material away

★ serves as a coolant to minimize the heat-affected zone (thereby preventing potential damage to the workpiece)

★ acts as a conductor for the current to pass between the electrode and the workpiece.

In wire EDM a very thin wire serves as the electrode. Special brass wires are typically used; the wire is slowly fed through the material and the electrical discharges actually cut the workpiece. Wire EDM is usually performed in a bath of water.

Answer the following questions:

1) What is EDM?

2) What is the difference between ram EDM and wire EDM?

3) How does EDM work?

Vocabulary Assistant

electrical discharge machining (EDM)电火花加工

conductive　传导性的,有传导力的,传导的

contour　轮廓(等高线,周线,电路,概要)

grinder　磨床;研磨机　　　　　　　hardened tool-steel　淬硬工具钢

titanium　钛　　　　　　　　　　carbide　碳化物,电石,硬质合金

spark 火花,电火花	flush away 冲去
successively 一个接一个地	crater 凹陷处,焊口
ram EDM 介质放电加工	wire EDM 丝电火花加工
graphite 石墨	attach 连接
submerge 使浸水,使陷入	fluid bath 液槽
current 电流;水流;气流	

1. The following are the advantages of EDM and benefits of wire EDM. Can you add any?

⚐ Advantages of EDM:

- very fragile parts can be machined
- the EDM process leaves no burrs
- the material is flushed away by the dielectric fluid
- it lowers costs by eliminating extra steps
- it reduces operating expenses, delivery dates
- it can replace many types of contour grinding operations
- eliminate secondary operations such as deburring and polishing
- it allows for cutting complex shapes without distortion

⚐ Benefits of wire EDM:

- efficient productive capabilities
- fast turnarounds
- reliable repeatability
- reducing costs
- burr free cutting
- tight tolerance and excellent finishes
- program files are downloadable
- accessing operation

Vocabulary Assistant

fragile　易碎的,脆的

burr　(金属切削或成型留下的)毛口,毛边

dielectric fluid　电介质　　eliminate　消除,剔除

grind　磨,碾　　　　　　　deburr　去除毛刺,修边

polish　磨光,磨料　　　　distortion　扭曲,变形

turnaround　转变,转盘　　tight tolerance　紧密度容限

program file　程序文件　　downloadable　可下载的

2. There are two primary EDM methods: ram EDM and wire EDM. Look at the following picture of wire EDM. Work with a partner to describe the working principle of wire EDM.

In wire EDM, a very thin wire serves as the electrode. Wire EDM is usually performed in a bath of water. In the following figure, A is the amplificatory scheme of the workpiece; B is the scheme of wire EDM working process. Here, 4 is used as the electrode to cut the workpiece. 7 helps to move in alternate directions. 3 is made as the supply of power.

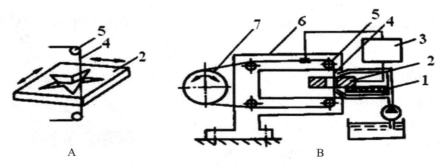

Working Principle of Wire EDM

1.绝缘底板　2.工件　3.脉冲电源　4.钼丝　5.导向轮　6.支架　7.储丝筒

Vocabulary Assistant

amplificatory 放大的	insulating base 绝缘底板
pulse power source 脉冲电源	molybdenum filament 钼丝
directive wheel 导向轮	bracket 支架

3. Look at the following four pictures. Do you know what kind of metals they are? And which can be used as electrodes for wire EDM?

A _____   B _____   C _____   D _____

- The materials that can be used for wire electrodes in wire EDM: _____.

4. Match the definitions with the terms.

A. electrode	B. voltage	C. discharge	D. frequency

_____ 1) electrical force measured in volts

_____ 2) rate of occurrence or repetition of sth, usually measured over a particular period of time

_____ 3) a conductor used to establish electrical contact with a nonmetallic part of a circuit

_____ 4) give or send out (liquid, gas, electric current, etc.)

Passage Reading

The Electrical Discharge Machining Process
电火花加工

1 It is known to every machinist that on standard machine tools electrical energy is converted into motion by an electric motor. Recently, we have discovered that electrical energy can be directly employed in metal removal. The main advantage is that metal of any hardness can be machined by electrical machining processes with good surface finish.

2 Electrical Discharge Machining (EDM) is a form of metal removal in which pulsating direct current is applied to a shaped tool (electrode) and a workpiece, both of which are capable of conducting electricity. The two are held close to each other with a non-conducting fluid serving as an insulator between them. When a voltage high enough to break down the insulator is reached, a spark jumps the gap between the tool and the workpiece. As a result, this spark removes a small portion of material.

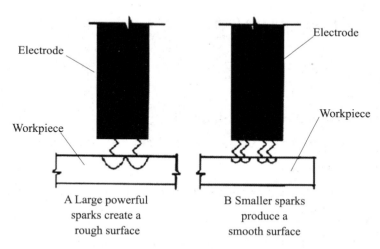

A Large powerful sparks create a rough surface

B Smaller sparks produce a smooth surface

The EDM Processes

Questions: *How does EDM work?*

What is the principle of EDM?

3 The EDM process is particularly advantageous in machining highly complicated shapes in forging, blanking, trimming, and other types of dies, and in machining hard and brittle materials. The major advantage of the process is that machining can be carried out in the hardened state of work materials, thus eliminating heat-treating and the possibility of subsequent cracking.

4 The other major advantage of the EDM process is that tools of softer material can be used to erode very hard and brittle materials. The tool materials generally used are copper, brass, aluminum, graphite, etc. Silver-tungsten and copper-tungsten are low-wearing tool materials, which may be used for finish machining. Cast iron can also be used during rough machining.

Questions: *Do you know in what process does EDM show its advantages?*

What are the major advantages of the EDM process?

5 The surface finish of a part produced on an EDM machine depends on the size of the sparks generated between it and the electrode. Big powerful sparks produce a rough surface and small, less powerful sparks produce a smooth surface. It is reported that metal removal rates and surface finish are controlled by the frequency and intensity of the spark. High frequency, low intensity sparks result in a low metal removal rate and produce a smooth finish; low frequency, high intensity sparks result in rapid metal corrosion and a coarse finish.

Questions: *What can control the metal removal rates and surface finish in an EDM process?*

What can determine the surface finish of a part produced on an EDM machine?

What kind of finish can high frequency, low-intensity sparks result in?

Vocabulary Assistant

surface finish　表面光洁度　　　　pulsating　脉动的,极为兴奋的
insulator　绝缘体　　　　　　　　portion　部分
graphite　石墨　　　　　　　　　silver-tungsten　钨银合金
copper-tungsten　钨铜合金　　　　finish machining　精加工
cast iron　铸铁　　　　　　　　　rough machining　粗加工
frequency　频率　　　　　　　　intensity　强度
coarse　（表面或质地）粗糙的

1. Fill in the table below by giving the corresponding translation.

English	Chinese
discharge energy	
	脆性材料
rib working	
	滚轮式电极
working allowance	
	旋转面
surface finish	
	管状电极
finish machining	
	粗加工

2. Find the definition in Column B which matches the words in Column A.

A	B
_____ 1) forge	a. person who operates a machine
_____ 2) blank	b. a light silvery metal, not tarnished by air
_____ 3) trim	c. find or learn about
_____ 4) brittle	d. make sth. smooth by cutting irregular parts
_____ 5) aluminum	e. erase sth.
_____ 6) machinist	f. shape sth. by heating in fire and hammering
_____ 7) discover	g. hard but easily broken; fragile

_____ 8) subsequent   h. of textures that are rough to the touch or substances consisting of relatively large particles

_____ 9) eliminate   i. following in time or order

_____ 10) coarse   j. do away with; take out

3. Use the correct form of the words from the box to complete the sentences.

> *remove   crack   subsequent   erode   metal   workpiece   current*

1) The _____ of phosphorus (磷) and sulfur (硫) requires special condition that can be met only by the basic process.

2) They made plans for a visit, but _____ difficulties with the car prevented it.

3) Don't pour hot water into the glass or it will _____.

4) Metals are _____ by acids.

5) Electrical Discharge Machining is a form of _____ removal.

6) In EDM, the pulsating direct _____ is applied to a shaped tool (electrode) and a _____, both of which are capable of conducting electricity.

4. Work with a partner to give brief answers to the following questions.

1) Describe the principle of EDM briefly.

_____

2) What can be used as tool materials in EDM?

_____

3) What materials are especially valuable in machining?

_____

4) What are the advantages of the EDM process?

_____

5) How can we control the metal removal rates and surface finish?

_____

5. Pair Activities: Ellen is an engineer. She is asked to explain EDM to the new students. (A: a new student, B: Ellen)

A: Hello, Ellen! Nice to see you again!

B: Hi! Me too!

A: Today, we have learned electrical discharge machining. I know that EDM can remove metal from a workpiece by using a series of electric sparks to erode material.

B: Right! That's it!

A: But I want to know more about EDM, for it's applied in so many fields.

B: OK! EDM is considered to be the most precision-oriented manufacturing process and is widely used for creating simple and complex shapes and geometries.

A: Yes!

B: Besides, EDM is favored in situations where high accuracy of work is required.

A: Then, could you please describe for me an electrical discharge machine?

B: No problem! It consists of a workpiece and the wire electrode. A workpiece is sometimes dipped in a dielectric to develop a potential difference between the workpiece and wire electrode. EDM works by eroding the material that appears in the electrical discharge path. This material is responsible for generating an arc between the workpiece and wire electrode. The wire electrode rotates cuts the internal cavities in the workpiece.

A: Oh, today our teacher have introduced to us wire EDM. Are there other types of EDM?

B: Actually, there are many types of electrical discharge machines. For example, a CNC EDM machine and a wire EDM machine. A CNC EDM machine is a computer numerical control machine and is used for removing metal using electrical discharge spark erosion. A wire EDM machine is designed for precision machining purposes and is used for cutting prismatic metal components. Other electrical discharge machines are commonly available.

A: Well, how does EDM function?

B: There are several ways in which electrical discharge machines function. Electrical discharge machines are used where fast turn around time is required, and they work by removing the workpiece that generates an arc with the wire electrode and creating a cavity in the workpiece. The dimensional accuracy required for an electrical discharge machine is + / − 0.0005 inches per inch. Furthermore, they are designed and manufactured to meet most industry requirements.

A: I've got it! Thank you, Ellen!

B: No thanks! Bye!

A: Bye-bye!

Answer the following questions:

1) What is an electrical discharge machine like according to the dialogue?

2) How many types of EDM machines are there? What are they?

3) How does EDM function?

6. Application

EDM is applied in many fields. Look at the following pictures and their corresponding explanations. Then, work with a partner and try to list some other applications of EDM.

A

EDM removes metal by means of spark erosion. The energy from the electrode dissipates the workpiece and the desired shape is formed. This principle has been widely applied in automobile in-dustry.

B

C

EDM's precision and intensity makes it produce any geometric shape. This leads to a standard method of producing prototypes and some production parts, particularly in low volume applications. Therefore, EDM technique is profoundly applied in producing medical tools.

EDM's absolute accuracy and repeatability can machine any tough material. So, it provides great cutting capability for aerospace industry. In this picture, people operate a total of 14 wire EDM systems to produce prototype and long-and-short-run production for the aerospace.

Vocabulary Assistant

dip　浸,蘸　　　　　　prismatic　棱镜的,柱状的
dissipate　消散,消失

Further Reading

UPDG and Its Applications
超精密金刚石磨削及其应用

Aspheric Surfaces

1 Ultra-precision machining based on single-point diamond turning and ultra-precision diamond grinding (UPDG) enables economic production of

optical, mechanical and electronic components or products with form accuracy and surface roughness in micrometre and nanometre ranges respectively. Associated form and surface roughness measurement technology also plays an important role in evaluating the quality of the ultra-precision machined surfaces. In Hong Kong, the ultra-precision machining centre was established to assist local industry with a full range of facilities and technical services for the rapid development and testing of high-quality ultra-precision products.

2 Hardened steel and brittle materials such as glass and ceramics are not normally amenable to diamond turning. Hardened steel contains high carbon content which reacts with the diamond and thus causes serious wear of the diamond tool. Brittle materials easily cause subsurface damages and chipping of the diamond tool.

3 UPDG employs a diamond wheel and requires a special machine of high precision and rigidity. It allows direct machining of brittle materials without the need for post polishing. It is particularly useful for aspheric optics production where ordinary polishing techniques is totally unsuitable.

4 The two basic precision requirements for ultra-precision diamond cutting tool are the precision of outline and edge radius. The crystal orientation and the crystal face selection of monocrystalline diamond are the basic knowledge of lapping

technology for ultra-precision diamond cutting tool. The grinding velocity, the grinding pressure, the grinding plate and grinding paste are the main technique factors that influence lapping efficiency and quality.

Applications of UPDG

Applications are now found in the production of:

⚐ Aspheric lenses in glass, fused silica or sapphire;

⚐ Complex surfaces made of ceramics;

▲ Ultra-precision moulds made in hardened steel or carbide alloys.

Vocabulary Assistant

aspheric　非球面的　　　　　　　　ultra-precision　超精密的

optical　光学的

micrometer　微米,百万分之一米(长度单位,符号为μm)

nanometer　纳米　　　　　　　　　amenable　受……影响的

chip　形成缺口;碎裂　　　　　　　optics　光学

edge radius　刃口半径　　　　　　　crystal orientation　晶体取向

crystal face　晶面　　　　　　　　　monocrystalline　单晶,单晶的

lap　研磨(玻璃,金属等)　　　　　　velocity　速度,速率

grinding plate　砂盘　　　　　　　　grinding paste　磨削用冷却剂

fused silica　熔融石英　　　　　　　sapphire　蓝宝石

carbide　电石;硬质合金

Over to you:

1) How is UPDG applied?

2) In what field is UPDG particularly applied?

3) Under what conditions is UPDG not properly used?

Unit 8

Electric Motors
电动机

Introduction

How does an electric motor work?

In any electric motor, operation is based on simple electromagnetism. A current-carrying conductor generates a magnetic field. When this is then placed in an external magnetic field, it will experience a force proportional to the current in the conductor, and to the strength of the external magnetic field. As we are well aware of from playing with magnets as a kid, opposite (North and South) polarities attract, while like polarities (North and North, South and South) repel. Fig. 8-1

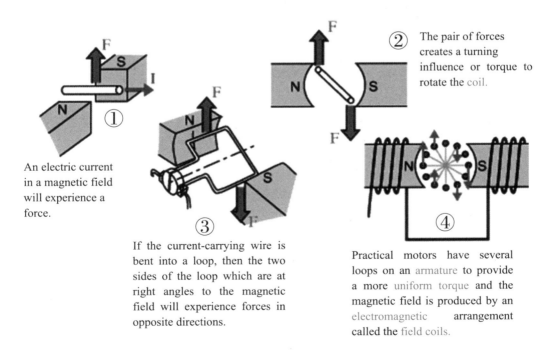

① An electric current in a magnetic field will experience a force.

② The pair of forces creates a turning influence or torque to rotate the coil.

③ If the current-carrying wire is bent into a loop, then the two sides of the loop which are at right angles to the magnetic field will experience forces in opposite directions.

④ Practical motors have several loops on an armature to provide a more uniform torque and the magnetic field is produced by an electromagnetic arrangement called the field coils.

Fig. 8-1 The Working Process of Electric Motor

shows the working process of electric motor.

The internal configuration of a DC motor is designed to harness the magnetic interaction between a current-carrying conductor and an external magnetic field to generate rotational motion. Every DC motor has six basic parts (Fig. 8-2)— axle, rotor, stator, commutator, field magnet(s), and brushes. In most common DC motors, the

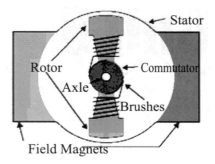

Fig. 8-2 DC motor

external magnetic field is produced by high-strength permanent magnets. The stator is the stationary part of the motor—this includes the motor casing, as well as two or more permanent magnet pole pieces. The rotor consists of windings (generally on a core), the windings being electrically connected to the commutator. Fig. 8-3 shows how DC motor works.

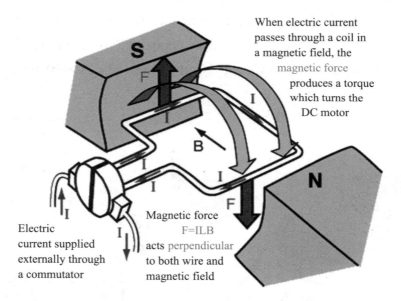

Fig. 8-3 The Working Process of DC Motor

AC Motor Operation

As in the DC motor case, a current is passed through the coil, generating a torque on the coil. Since the current is alternating, the motor will run smoothly only at the frequency of the sine wave. It is called a synchronous motor. More common is the induction motor, where electric current is induced in the rotating coils rather than supplied to them directly.

One of the drawbacks of this kind of AC motor is the high current which must flow through the rotating contacts. Sparking and heating at those contacts can waste energy and shorten the lifetime of the motor. In common AC motors, the magnetic field is produced by an electromagnet powered by the same AC voltage as the motor coil. The coils which produce the magnetic field are

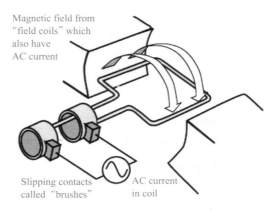

Fig. 8-4 The Working Process of AC Motor

sometimes referred to as the "stator", while the coils and the solid core which rotates is called the "armature" (see Fig. 8-4).

Now look at Fig. 8-5. Can you identify which one is AC motor, and which one is DC motor? And discuss the function of each one with a partner.

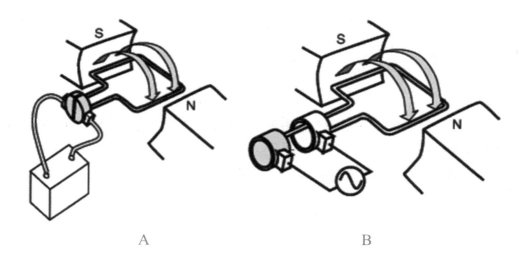

A   B

Fig. 8-5

Vocabulary Assistant

electromagnetism 电磁,电磁学   external magnetic field 外磁场

polarity 磁性引力,极性   coil 线圈,卷材

armature 电枢

electromagnetic 电磁铁的,电磁体的   uniform torque 不变的扭动力

internal configuration 内部构形   field coil 励磁线圈

harness 利用产生动力   DC motor 直流电动机

rotational motion 回转运动   interaction 相互作用,相互影响

rotor 转子   axle 轮轴,车轴

commutator 整流器   stator 固定子,固定片

permanent magnet 永磁体   field magnet 场磁体

stationary 静止不动的;固定的,稳定的

casing 机壳   winding 绕组,绕圈

magnetic force 磁力   perpendicular 垂直的

AC motor 交流电动机   sine wave 正弦波

synchronous motor 同步电动机   induction motor 感应电动机

induce 产生,带来   sparking 冒火花,产生火花

1. The following are the terms of different types of motors. Do you know them? Can you add any?

1. squirrel-cage motor 双鼠笼式电动机

2. box-frame motor 箱形机座电动机

3. canned motor 封闭电动机;密封式发动机

4. capacitor induction motor 电容电动机

5. capacitor split-phase motor 电容分相式电动机

6. ceramic permanent-magnet motor 陶瓷永磁电动机

7. close-ratio two-speed motor 近比率双速电动机

8. compensated induction motor　补偿式感应电动机

9. condenser shunt type induction motor　电容分相式感应电动机

10. consequent-poles motor　变极式双速电动机;交替磁极式电动机

2. The following is a picture of an electric motor. Can you name each part with the help of the given terms? Then work with a partner to match them.

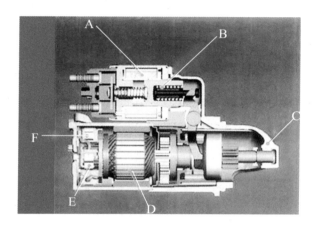

1. long-life brush — _____

2. antifreezing coil — _____

3. high-intensity molding commutator — _____

4. anti-high impact end closure — _____

5. equipment of monolithic movable core contact small plate — _____

6. ferrite permanent magnet — _____

Vocabulary Assistant

antifreezing　防冻的	high-intensity　高强度的
high impact　高冲击	monolithic　整体式的
movable core　动铁芯	contact small plate　接触片
ferrite　铁氧体,铁素体	

機電英語(第二版)

3. Match the definitions with the terms.

A. current   B. commutator   C. magnet   D. insulate

_____ 1) protect sth. by covering it with a material that prevents sth. (esp. heat, electricity or sound) from passing through

_____ 2) piece of iron, often in a horseshoe shape, which can attract iron, either naturally or because of an electric current passed through it, and which points to roughly north and south when freely suspended

_____ 3) device for altering the direction of an electric current

_____ 4) movement of water, air, etc. flowing in a certain direction through slower-moving or still water or air; flow of electricity through sth. or along a wire or cable

Passage Reading

Electric Motors
电动机

1 Each type of motor has its particular field of usefulness. Because of its simplicity, economy, and durability, the induction motor is more widely used for industrial purposes than any other type of AC motor, especially if a high-speed drive is desired.

2 If AC power is available, all drives requiring constant speed should use squirrel-cage induction or synchronous motors because of their ruggedness and lower cost. Drives requiring varying speeds, such as fans, blowers, or pumps, may be driven by wound-rotor induction motors. However, if there are machine tools or other machines requiring adjustable speed or a wide range of speed control, it will probably be desirable to install DC motors on such machines to supply them with electric power generated from the AC system by motor-generator sets or electronic rectifiers.

Questions: *What kind of motor is most widely used for industrial purposes? Why?*
On what occasion will DC motors be installed?

3 Almost all constant-speed machines may be driven by AC squirrel-cage motors because these motors are made with a variety of speed and torque characteristics. When power supply is limited, the wound-rotor motor is used, even to drive constant-speed machines.

4 For varying-speed service, wound-rotor motors with resistance control are

 used for fans, blowers, and other apparatus for continuous duty and are used for cranes, hoists, and other installations for intermittent duty. The controller and resistors must be properly chosen for the specific application. Synchronous motors may be used for almost any constant-speed drive requiring about 100 hp or over.

Questions: *Why are AC squirrel-cage motors used to drive almost all constant-speed machines?*
When power supply is limited, what kind of motor is used?

5 Cost is an important factor when more than one type of AC motor is applicable. The squirrel-cage motor is the least expensive AC motor of the three types, and requires very little control equipment. The wound-rotor is more expensive and requires additional secondary control. The synchronous motor is even more expensive and requires a source of DC excitation, as well as special synchronizing control to apply the DC power at the correct instant. When very large machines are involved, for example, 1,000 hp or over, the cost price may change considerably and should be checked on an individual basis.

6 The various types of single-phase AC motors and universal motors are used very little in industrial applications, since poly-phase AC or DC power is generally available. When such motors are used, they are usually built into the equipment by the machinery manufacturer, as in portable tools, office

machinery, and other equipment. These motors are, as a rule, especially designed for the specific machines with which they are used.

Questions: *What factor should be considered first when more than one type of AC motor is applicable?*
What kind of motor is the least expensive one?
Why are single phase AC motors and universal motors used very little?

Vocabulary Assistant

induction motor 感应电动机
constant speed 恒速
fan 电风扇
wound-rotor induction motor 线绕式转子感应电动机
adjustable 可调整的,可控制的
motor-generator set 电动发电机设备
rectifier 整流器
crane 起重机
intermittent 间歇的,断断续续的
hp (horse power) 马力
excitation 励磁
universal motor 交直流两用电动机
as a rule 通常

high-speed drive 高速驱动
ruggedness 坚固性
blower 鼓风机

apparatus 仪器,装置
hoist 卷扬机,升降机
controller 控制器
control equipment 控制装置
single-phase 单相
poly-phase 多相,多相的

1. Fill in the table below by giving the corresponding translation.

English	Chinese
low cost	
	变速
speed control	
	电源
speed and torque characteristics	
	持续作业
intermittent duty	
	特殊应用
single-phase	
	便携式工具

2. Use the correct form of the words from the box to complete the following sentences.

| ruggedness   control   induction   aware   low cost   regular   overcome |

1) The firm has to _____ its resistance to new technology.

2) The manager is fully _____ of the cost price in the factory.

3) The squirrel-cage motors require very little _____ equipment.

4) Squirrel-cage motors are known for their _____ and _____.

5) The _____ motor is usually used especially when a high-speed drive is desired.

6) The motors should be checked on a _____ basis.

3. Find the definition in Column B which matches the words in Column A.

A	B
_____ 1) machinery	a. belonging to, done by all; affecting all
_____ 2) universal	b. machines in general
_____ 3) rectifier	c. coming and going at intervals; not continuous
_____ 4) intermittent	d. being likely to last for a long time
_____ 5) pump	e. machine or device for forcing liquid, gas or air into, out of or through sth.

_____ 6) durability   f. force that hinders or stops sth.

_____ 7) application   g. something that attracts

_____ 8) resistance   h. act of putting a theory, discovery, etc. to practical use

_____ 9) rotate   i. a device for converting alternating current into direct current

_____ 10) magnet   j. (cause sth. to) move in circles round a central point

4. Work with a partner to give brief answers to the following questions.

1) What are the advantages of the induction motor?

_____

2) Why should all drives requiring constant speed use squirrel-cage induction or synchronous motors if AC power is available?

_____

3) When is the induction motor most useful?

_____

4) How many types of motors are discussed in this passage?

_____

5) Why are the various types of single-phase AC motors and universal motors seldom used in industrial applications?

_____

5. Pair Activities: Steven is a motor seller. He is asked to introduce a kind of new product of electric motor to customers. (A: customer; B: Steven)

B: Today, I'll introduce to you a new product. Its name is Y series (IP44) TEFC squirrel-cage three phase asyn-chronous motor.

A: What's that? Does it have some spe-cial characteristics?

B: Of course! Y series (IP44) motor is a series of product designed in our country. It is of light weight, good

performance of electricity, advanced economic target and secure structure, simple operation, and easy maintaining. It is the most advanced motor made in China which can be the equipment driven by electricity widely and used for agriculture, factories mines and enterprises.

A: It sounds good! But could you please make it clearer, for example, how about its structure, or functions?

B: No problem! Y series (IP44) motor is fully enclosed, fan-cooled three phase squirrel-cage asynchronous motor. The protection degree IP44 for case is adopted. Its structure can prevent dust, iron filings or other splashy object from intruding into motor. Technically speaking, there are three types:

★ B3—Frame with foot, end shield without flange.

★ B5—Frame without foot, end shield with flange, dimension of which is greater than frame.

★ B35—Frame with foot, end shield with flange, dimension of which is greater than frame.

A: I want to know something more about its technical working condition.

B: Its rated voltage is 380V; rated frequency is 50Hz; protection class is IP44; its insulation class is B class. Moreover, Y series motors are commonly designed for operation at an ambient temperature of not exceeding +40℃. An altitude not over 1000m above sea level and a relative humidity air not higher than 95%.

A: Oh, then in what specific field can we apply it?

B: Ok, I'm just going to mention that. Actually, Y series (IP44) motor may be used for driving the machinery without any special requirement in the rotational speed and performance, such as metal-cutting machines, pumps,

blowers, mining machines, agriculture machines etc. As a result of better starting performance, the motors are also suitable for use in machinery that require a higher starting torque, for example, compressors, mixers and conveyors etc.

A: Really! Fantastic! Well, let's talk about the price!

Answer the following questions:

1) What kind of electric motor is Y series (IP44) motor? And in what field can we apply it?

2) Can you describe the technical working conditions of Y series (IP44) motor? And what are they?

3) How many types of Y series (IP44) motor are there?

Vocabulary Assistant

secure 牢固	factory mine 工矿企业

fan-cooled three phase squirrel-cage asynchronous motor 扇冷式三相异步电动机

iron filing 铁屑	splashy object 飞溅物体
intrude 入侵	end shield 端盖
flange 凸缘	technical condition 技术条件
rated voltage 额定电压	rated frequency 额定频率
protection class 防护等级	insulation class 绝缘等级
ambient 环境的,外界的	exceed 超过
altitude 海拔,标高	sea level 海平面
relative humidity 相对湿度	rotational speed 转动速度,周围速度
metal-cutting machine 金属切削机械	
mining machine 采掘机	compressor 压缩机
mixer 混合机,搅拌机	

Further Reading

Servo Motors
伺服电动机

Servo motors are used in closed loop control systems (Fig. 8-6). The digital servo motor controller directs operation of the servo motor by sending velocity command signals to the amplifier, which drives the servo motor. An integral feedback device or devices (encoder and tachometer) are either incorporated within the servo motor or are remotely mounted, often on the load itself. These provide the servo motor's position and velocity feedback that the controller compares to its programmed motion profile and uses to alter its velocity signal. Servo motors feature a motion

profile, which is a set of instructions programmed into the controller that defines the servo motor operation in terms of time, position, and velocity. The ability of the servo motor to adjust to differences between the motion profile and feedback signals depends greatly upon the type of controls and servo motors used.

Three basic types of servo motors are used in modern servo systems: AC servo motors, based on induction motor designs; DC servo motors, based on DC motor designs; and AC brushless servo motors, based on synchronous motor designs.

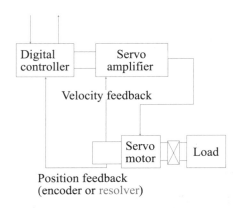

Fig. 8-6 Typical DC servo motor system with either encoder or resolver feedback.
Some older servo motor systems use a tachometer and encoder for feedback.

DC Servo Motors

DC servo motors are normally used as prime movers in computers, numerically controlled machinery, or other applications where starts and stops are made quickly and accurately. Servo motors have lightweight, low-inertia armatures that respond quickly to excitation-voltage changes. In addition, very low armature inductance in these servo motors results in a low electrical time constant (typically 0.05 to 1.5 msec) that further sharpens servo motor response to command signals. Servo motors include permanent-magnetic, printed-circuit, and moving-coil. The rotor of a coil DC servo motor consists of a cylindrical coil of copper or aluminum wire coils which rotate in a magnetic field in the annular space between magnetic pole pieces and a stationary iron core. The servo motor features a field,

which is provided by cast alnico magnets whose magnetic axis is radial. Servo motors usually have two, four, or six poles.

DC servo motor characteristics include inertia, physical shape, costs, shaft resonance, shaft configuration, speed, and weight. Although these DC servo motors have similar torque ratings, their physical and electrical constants vary.

AC Servo Motors

AC servo motors are used in applications requiring rapid and accurate response characteristics. To achieve these characteristics, these AC servo motors have small diameter, high resistance rotors. The AC servo motor's small diameter provides low inertia for fast starts, stops, and reversals. High resistance provides nearly linear speed-torque characteristics for accurate servo motor control.

An induction motor designed for servo use is wound with two phases physically at right angles or in space quadrature. A fixed or reference winding is excited by a fixed voltage source, while the control winding is excited by an adjustable or variable control voltage, usually from a servo amplifier. The servo motor windings are often designed with the same voltage, so that power inputs at maximum fixed phase excitation, and at maximum control phase signal, are in balance. The inherent damping of servo motors decreases as ratings increase, and the servo motors are designed to have a reasonable efficiency at the sacrifice of speed-torque linearity.

Vocabulary Assistant

servo motor　伺服电动机

closed loop control system　闭合环路式控制系统;闭环控制系统

velocity command signal　速度控制信号

amplifier　放大器,扩音机　　integral　构成整体所必需的,完整的

encoder　编码器　　　　tachometer　转速计

remotely　远程地　　　programmed motion profile　程序运动轮廓

resolver　分解器　　　prime　首要的,重要的

low-inertia　低惯性的　　inductance　感应系数,自感应

msec=megasecond　兆秒　　printed circuit　印刷电路

moving-coil　动圈式　　　cylindrical coil　圆柱形线圈

annular　环状的　　　　alnico magnet　铝镍钴磁钢

radial　放射(式)的,光线的　inertia　惯性,惯量

shaft resonance　轴共振　　shaft configuration　轴结构

high resistance　高阻的　　linear　直线的,线状的

speed-torque　转速力矩　　space quadrature　空间90度相位差

damping　阻尼　　　　linearity　线性,直线性

Over to you:

1) How is servo motor used?

2) How many types of servo motors are used in modern servo systems?

3) How are AC and DC servo motors used?

Unit 9

Industrial Robots
工业机器人

Introduction

Industrial robots are not like the science fiction devices that possess human-like abilities and provide companionship to space travelers. Research to enable robots to "see", "hear", "touch", and "listen" has been underway for two decades and is beginning to bear fruits. However, the current technology of industrial robots is such that most robots contain only an arm rather than all the anatomy a human possesses. Current control only allows these devices to move from point to point in space, performing relatively simple tasks.

The typical structure of industrial robots consists of four major components (Fig. 9-1): the manipulator, the end effector, the power supply, and the control system.

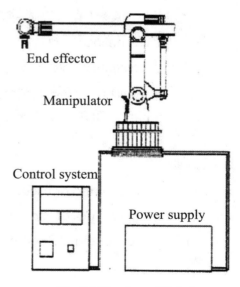

Fig. 9-1 Structure of Robots

■ **The manipulator** is a mechanical unit that provides motions similar to those of a human arm. It often has a shoulder joint, an elbow and a wrist. It can rotate or slide, stretch out and withdraw in every possible direction with certain flexibility.

■ **The effector** attaches itself to the end of the robot wrist, also called end-of-arm tooling. It is a device intended for performing the designed operations as a human hand can. End effectors are generally custom-made to meet special handling requirements.

■ **The power supply** is the actuator for moving the robot arm, controlling the joints and operating the end effector. The basic types of power sources include electrical, pneumatic, and hydraulic.

■ **The control system** is the commutations and information-processing system that gives commands for movements of the robot. It is the brain and nerves of the robot. There are open-loop controller and close-loop controller. The open-loop controller controls the robot only by following the predetermined step-by-step instructions. The close-loop controller system use feedback sensors to produce signals that reflect the current state of the controlled objects.

Industrial robots vary widely in size, shape, number of axes, degrees of freedom, and design configuration. Each factor influences the dimensions of the robot's working envelope or the volume of space within which it can move and perform its designed task. A broader classification of robots can be described as below.

★ **Fixed and Variable-Sequence Robot.** The fixed-sequence robot (also called a pick-and-place robot) is programmed for a specific sequence of operations. Its movements are from point to point, and the cycle is repeated continuously. The variable-sequence robot can either be programmed for a specific sequence of operations or be reprogrammed to perform another sequence of operation.

★ **Playback Robot.** An operator leads or walks the playback robot and its end effector through the desired path. The robot memorizes and records the path and sequence of motions and can repeat them continually without any further actions or guidance by the operator.

★ **Numerically Controlled Robot.** The numerically controlled robot is programmed and operated much like a numerically controlled machine. The robot is servocontrolled by digital data, and its sequence of movements can be changed with relative ease.

★ **Intelligent Robot.** The intelligent robot is capable of performing some of the functions and tasks carried out by human beings. It is equipped with a variety of sensors with visual and tactile capabilities.

1. The following is the typical structure of an industrial robot. Can you match the terms with each part?

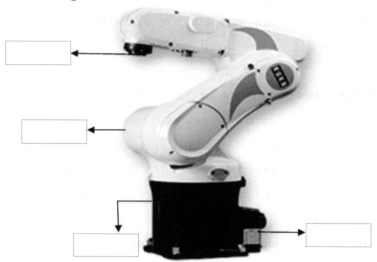

① manipulator ② end effector ③ power supply ④ control system

Vocabulary Assistant

companionship 交谊,友谊	anatomy 剖析,解剖
possess 占有,拥有	manipulator 操作者,操纵器
effector 受动器,效应器,操纵装置	
shoulder joint 肩关节	elbow 肘
wrist 腕,机械腕	stretch out 伸
withdraw 缩回,撤出	actuator 传动机构,驱动器
pneumatic 气力的	hydraulic 水力的,液压的
commutation 交换	open-loop 开环,开放式回路
close-loop 闭环,封闭式回路	sensor 传感器

configuration	构造, 结构	fixed-sequence robot	固定顺序机器人
sequence	顺序, 次序	continuously	连续地
operator	操作者	playback robot	再生式机器人
variable-sequence robot	可变顺序机器人		
motion	运动, 动作	servocontrol	伺服控制
numerically controlled robot	数控机器人		
intelligent robot	智能机器人	visual	视觉的
tactile	触觉的	capability	能力

Answer the following questions:

1) Can industrial robots think and speak?

2) What is the typical structure of an industrial robot?

3) What is control system? Describe the two types of the control system.

2. The following is the description of differences and similarities between industrial robots and NC (numerical control machine tool). Can you add any?

Similarities

◆ power drive (electrical motor, hydraulic system, pneumatic system)

◆ controllers (open-loop or close-loop)

◆ feedback system

◆ some of the industrial applications

Differences

◆ The robot is a lighter, more portable piece of equipment

◆ The applications of robot are more general

◆ Programming

　　—NC programming has been performed off-line

　　—Robot programming has been performed on-line

3. Here are the major categories of industrial robots. Work with a partner to discuss functions of each.

Typical industrial robots do jobs that are difficult, dangerous or dull. They lift heavy objects, paint, handle chemicals, and perform assembly work. They perform the same job hour after hour, day after day with precision. They won't get tired or make errors associated with fatigue, so they are ideally suited to performing repetitive tasks. The major categories of industrial robot by mechanical structure are:

Cartesian robot/Gantry robot (Fig. 9-2) is used for pick-and-place-work, application of sealant, assembly operations, handling machine tools and arc welding. It's a robot whose arm has three prismatic joints, whose axes are coincident with a Cartesian coordinator.

Fig. 9-2 Cartesian Robot/ Gantry Robot

Cylindrical robot (Fig. 9-3) is used for assembly operations, handling machine tools, spot welding, and handling die casting machines. It's a robot whose axes form a cylindrical coordinate system.

Fig. 9-3 Cylindrical Robot

Spherical robot/Polar robot (Fig. 9-4) is used for handling machine tools, spot welding, die casting, fettling machines, fettling and arc welding. It's a robot whose axes form a polar coordinate system.

Fig. 9-4 Spherical Robot/Polar Robot

SCARA robot (Fig. 9-5) is used for pick-and-place work, application of sealant, assembly operations and handling machine tools. It's a robot which has two parallel rotary joints to provide compliance in a plane.

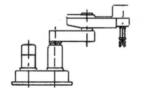

Fig. 9-5 SCARA Robot

Articulated robot (Fig. 9-6) is used for assembly operations, die casting, fettling machines, gas welding, arc welding and spray painting. It's a robot whose arm has at least three rotary joints.

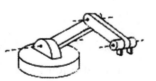

Fig. 9-6 Articulated Robot

Rectangular coordinate robot (Fig. 9-7) is used for handling cockpit flight simulators with a mobile platform. It's a robot whose arms have concurrent prismatic or rotary joints.

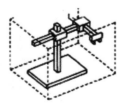

Fig. 9-7 Rectangular Coordinale Robot

Vocabulary Assistant

precision	精确, 精密度	fatigue	劳累
repetitive	反复的	cartesian robot	直角坐标机器人
gantry	构台, 桶架	gantry robot	拱架机器人
sealant	密封剂	joint	关节, 结合点
coincident	一致的, 符合的	cylindrical robot	圆柱坐标机器人
spot welding	定位焊接	spherical	球形的
spherical robot	环坐标机器人	polar	极性的
polar robot	极坐标机器人	fettle	修整, 修整
polar coordinate	极坐标		

SCARA (selective compliance assembly robot arm)　平面关节机器人

parallel	平行的, 相同的	rotary	旋转的, 循环的
compliance	依从, 顺应	articulated robot	关节型机器人
spray painting	喷漆	rectangular	矩形的, 成直角的
cockpit	驾驶员座舱	simulator	模拟器
platform	平台, 月台, 讲台	concurrent	协作的

4. Look at the following classifications of industrial robots. Then work with a partner to give the English version to the following classifications.

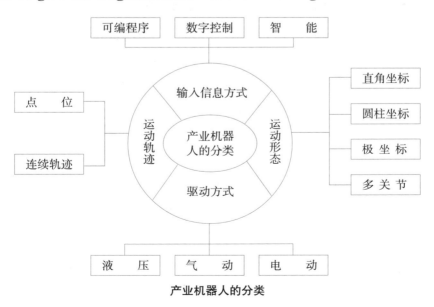

产业机器人的分类

5. Match the definitions with the terms.

| A. *hydraulic* | B. *rectangular* | C. *actuator* | D. *sensor* |

_____ 1) a mechanical device for moving or controlling something

_____ 2) a device that responds to a physical stimulus and transmits a resulting impulse

_____ 3) operated by the resistance offered or the pressure transmitted when a quantity of liquid (as water or oil) is forced through a tube

_____ 4) crossing, lying, or meeting at a right angle

Passage Reading

Industrial Robots
工业机器人

1 An industrial robot is a general-purpose, programmable machine possessing certain humanlike characteristics. The most typical humanlike characteristic of the robot is its arm. This arm, together with the robot's capacity to be programmed, makes it ideally suited to a variety of production tasks, including machine loading, spot welding, spray painting, and assembly. The robot can be programmed to perform a sequence of mechanical motions, and it can repeat those motions over and over again until reprogrammed to perform some other jobs.

Questions: *What is an industrial robot?*

What can an industrial robot do?

2 An industrial robot shares many attributes in common with a numerical control machine tool. NC technology used to operate machine tools is also used to actuate the robot's mechanical arm. The robot is a lighter, more portable piece of equipment than a NC mechanical tool. The applications of the robot are more general, typically involving the handling of workparts. Also, the programming of the robot is different from NC part programming. Traditionally, NC programming has been performed off-line with the machine commands being contained on a punched tape. Robot programming has usually been accomplished on-line, with the instructions being retained in the robot's electronic memory. In spite of these differences, there are definite similarities between robots and NC machines in terms of power drive technologies, feedback systems, the trend toward computer control, and even some of the industrial applications.

3 The popular concept of a robot has been fashioned by science fiction novels and movies such as "Star Wars". These images tend to exaggerate the robot's similarity to human anatomy and behavior. The human analogy has sometimes been a troublesome issue in industry. People tend to associate the future use of advanced robots in factories with high unemployment and the conquer of human beings by these machines.

Questions: *What does NC refer to?*

What is the popular concept of a robot?

4 Largely in response to this humanoid conception associated with robots, there have been attempts to develop definitions which reduce the humanlike impact. The Robot Institute of America has developed the following definition:

A robot is a programmable, multi-function manipulator designed to move material, parts, tools, or special devices through variable programmed motions for the performance of a variety of tasks.

5 Attempts have even been made to rename the robot. George Devol, one of the original inventors in robotics technology, called his patent application by the name "programmed article transfer." For many years, the Ford Motor Company used the term "universal transfer device" instead of "robot." Today the term "robot" seems to have become entrenched in the language, together with whatever humanlike characteristics people have attached to the device.

6 The application in which robots are used are quite broad. These applications can be grouped into three categories: material processing, material handling and assembly.

7 In material processing, robots use tools to process the raw material. For example, the robot tools could include a drill and the robot would be able to perform drilling operations on raw material.

8 Material handling consists of loading, unloading, and transferring of workpieces in manufacturing facilities. These operations can be performed reliably and repeatedly with robots, thereby improving quality and reducing scrap losses.

9 Assembly is another large application area for using robotics. An automatic assembly system can incorporate automatic testing, robot automation and mechanical handling for reducing labor.

Questions: *Why should robots be redefined?*

Why does the Ford Motor Company use the term "universal transfer device" instead of "robot"?

Vocabulary Assistant

general-purpose	多种用途的	characteristic	特征,特性
capacity	能力,容量,电容	machine loading	机器装载,机器负荷
sequence	次序,顺序	reprogram	改编,程序重调

attribute　属性,品质　　　　actuate　开动(机械等);驱使

off-line　脱机,离线　　　　punched tape　穿孔纸带

fashion　将……做成形状,设计,制造　　analogy　类似,模拟

conquer　占领　　　　humanoid　具有人的特点的

multi-function manipulator　多功能操纵器

patent　专利　　　　transfer　转移

entrenched　确立的　　　category　种类,范畴

material processing　材料加工　　drill　钻孔

loading　装载　　　　unload　卸载

reliably　可靠地　　　scrap　废料

test　测试

1. Fill in the table below by giving the corresponding translation.

English	Chinese
feedback	
	液压系统
welding	
	装配
effector	
	装载
general-purpose	
	定位焊接
cylindrical	
	废料

2. Find the definition in Column B which matches the words in Column A.

A	B
_____ 1) assembly	a. person who handles things manually
_____ 2) conquer	b. following of one thing after another in time
_____ 3) sequence	c. parts of machine, vehicle, etc. put together
_____ 4) programming	d. take possession of by force,
_____ 5) command	e. setting an order and time for planned events
_____ 6) original	f. a piece of equipment intended for a particular purpose

_____ 7) device   g. an order that should be obeyed

_____ 8) instruction   h. preceding all others in time

_____ 9) manipulator   i. the point of connection

_____ 10) joint   j. message describing how something is to be done

3. Use the correct form of the words from the box to complete the sentences.

| auto | program | industry | robot | application | manipulate |

1) Because robots can perform certain basic tasks more quickly and accurately than humans, they are being increasingly used in various manufacturing _____.

2) The future developments of _____ depend mostly on the young scientists, who are less conservative, who have active and imaginative brains and not learned to think "not practical", or "not possible".

3) Robots can work 24 hours a day for years on end with no failures whatsoever. In many applications (particularly those in the auto industry) they are _____ once and then repeat that exact same task for years.

4) The robot is a very special type of production tool. As a result, the _____ in which robot are used are quite broad.

5) The _____ industry is a major user of robotic spot welders.

6) A robot is a programmable, multi-functional _____ designed to move material, parts, tools, or special devices through variable programmed motions for the performance of a variety of tasks.

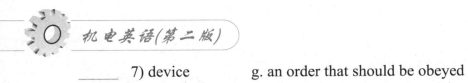

4. Pair Activities: Steven is a student majoring in Mechanical and Electrical Engineering. He is asked to explain the term industrial robot to a new student. (A: new student, B: Steven)

A: Do you know industrial robot?

B: Yes, industrial robot is an automatically controlled, reprogrammable, multi-purpose, manipulating machine with several reprogrammable axes, which may be either fixed in place or mobile for use in industrial automation application.

A:　When was robot first used?

B:　As for the term "robot", it was first used in a play called "R.U.R." or "Rossum's Universal Robots" by the former Czech writer Karel Capek in 1921.

A:　Then when was the industrial robot invented?

B:　The industrial robot was first invented in 1956 by two Americans: George Devol and Joseph Engelberger, and they set up the first robot company in the world.

A:　I can hardly imagine how fast the robotic technology develops for it is so widely used around us.

B:　Of course. With the development of technology, the industrial robots have been used in various fields, for example, manufacturing industry, mechanical processing industry, electrical and electronic industry, rubber and plastic industry, food industry, wood manufacturing industry and so on. Here I have a chart of the major sectors of the industrial robot demand distribution ratio in America in 2005. You can find how it is used in each field.

A:　So, why are robots so useful in industry?

B:　There are a variety of reasons. You know robots often make those business owners more competitive, because robots can do things more efficiently than humans. For instance, robots never get sick or need to rest, so they can work 24 hours a day, 7 days a week; or when the task required would be dangerous for a person, robots can do the work instead. What's more, robots don't get bored, so no work will be a problem for a robot.

A:　But robots can't do every type of job.

B:　Sure. However, there are certain industrial tasks robots do very well including: First, assembling operations, which accounts for about 33% of the applications of the world robot stock (1997). Many of these robots can be found in automobile and electronic

industry. Second, continuous arc welding and spot welding is one of the most common uses for industrial robots. Nearly 25% of all industrial robots are used in different welding applications. Third, packaging/ palletizing, which is still a minor application area for industrial robots, accounts for only 2.8% (1997). Besides, there are other industrial tasks like material removal, machine loading, material transfer, cutting operations, part inspection, part sorting, part cleaning, and part polishing. The application area is expected to grow as robots become easier to handle.

A: Great! Thank you for your explanation. See you later!

B: See you!

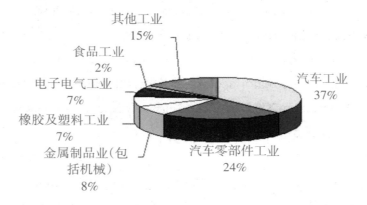

Fig. 9-8 2005年美洲地区各主要行业对工业机器人需求比例分布图

Vocabulary Assistant

multipurpose　多种用途的　　　　　mobile　可移动的
Czech　捷克人的　　　　　　　　　　ratio　比率
competitive　竞争的,有竞争力的　　assembling operation　组装业务
account for　（在数量比例方面）占　stock　存货

palletizing	夹板装载	minor	次要的
sort	整理, 分类	polish	磨光

> **Answer the following questions:**
>
> *1. Can you describe the development of robots?*
>
> *2. Do you know in what field industrial robot can be applied?*

5. Application

Nowadays, the industrial robots are used in various fields. Look at the following pictures and decide what each of them refers to. Find correct explanations for each of them.

A

B

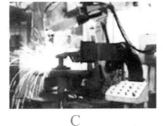

C

D

_____ 1) An industrial robot is used in welding.

_____ 2) An industrial robot is used in motor manufacturing industry.

_____ 3) An industrial robot is used in distribution.

_____ 4) An industrial robot is used in underwater work.

Over to you:

True or False Questions

_____ 1) Robots can repeat the motion sequence over and over again and then stop by themselves.

_____ 2) Compared with a robot, the uses of an NC machine tool are more general.

_____ 3) Traditionally, NC programming has been performed on-line with the machine commands being contained on a punched tape.

_____ 4) There are definite similarities between robots and NC machines, such as feedback systems.

_____ 5) Science fiction novels and movies exaggerate the robot's similarity to human anatomy and behavior.

Further Reading

Mechatronics
机电一体化

Mechatronics was originally coined in 1970s from the integration of two engineering disciplines—mechanics and electronics. More recently, with advancements in the areas of control and communications, the word mechatronics has been adapted as the synergetic integration of three disciplines: mechanics, control and electronics and is aimed at the study of mainly manufacturing machines controlled by electronics. It is also being viewed as the fusion of mechanical engineering with electronics  and intelligent computer control in the design and manufacture of industrial products and processes. They are called "intelligent" or "smart" machines because they are associated with "intelligence" or "smartness". Mechatronics is also defined as the synergetic integration of mechanical engineering with electronics and intelligent computer control in the design and manufacture of products and processes.

In engineering terms, what can be made to emerge is a new and previously unattainable set of performance characteristics. Thus mechatronics is truly an interdisciplinary subject drawn from mechanical, electrical, electronics, computer, and manufacturing engineering. The technology areas of mechatronics involve system modeling, simulation, sensors and measurement systems, drive and actuation system, analysis of the behavior of systems, control systems and microprocessor systems.

Mechatronic Paradigm

Mechatronic paradigm deals with benchmarking and emerging problems in engineering, science, and technologies which have not been solved. Mechatronics is an integrated comprehensive study of intelligent and high performance electromechanical system (mechanisms and processes) and motion control through the use of advanced microprocessors and DSPs, power electronics and ICs, design and optimization, modeling and simulation, analysis and virtual prototyping, etc. Mechatronics integrates electrical, mechanical, and computer engineering areas (see Fig. 9-9). Thus it has the integrated multidisciplinary features.

One of the most challenging problems in mechatronics systems design is the development of system architecture, e.g., selection of hardware (actuators, sensors, devices, ICs, microcontrollers, and DSPs) and software (environment and computation algorithms to perform sensing and control, information flow and data acquisition, simulation, visualization, and virtual prototyping). Attempts to design state-of-the-art

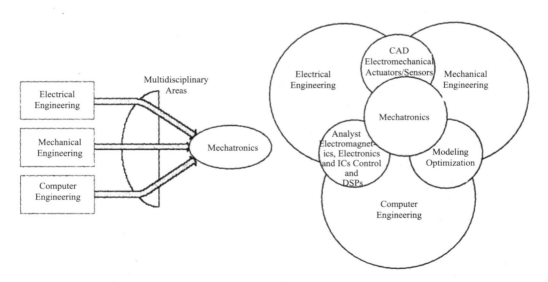

Fig. 9-9 Mechatronics Integrates Electrical, Mechanical and Computer Engineering

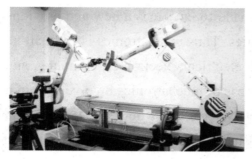

man-made mechatronic systems and to guarantee the integrated design can be pursued through analysis of complex patterns and paradigms of evolutionary developed biological systems. Recent trends in engineering have increased the emphasis on integrated analysis, design and control of advanced mehanical systems. The scope of mechantronic system has continued to expand, and in addition to actuators, sensors, power electronics, ICs microprocessors, DSPs, as well as input/output devices, many other subsystems must be integrated, even though the basic foundations have been developed, some urgent areas have been downgraded, less emphasized and covered. To overcome these difficulties, the mechatronic paradigm was introduced with the ultimate goals to:

(1) guarantee an eventual consensus and ensure descriptive multidisciplinary features;

(2) extend and augment the results of classical mechanics, electromechanical systems, power electronics, ICs, and control theory to advanced hardware and software;

(3) acquire and expand the engineering core integrating interdisciplinary areas;

(4) link and place the integrated perspectives of electromechanical systems, power electronics, ICs, DSPs, control, signal processing, MEMS and NEMS in the

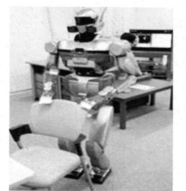

engineering curriculum in favor of the common structure needed.

The study of high-performance electromechanical systems should be considered as the unified cornerstone of the engineering curriculum through mechatronics. The unified analysis of actuators and sensors (e.g. electromechanical motion devices), power electronics and ICs, microprocessors and DSPs, advanced hardware and software, has barely been introduced in the engineering curriculum. Mechatronics, as a breakthrough concept in design and analysis of conventional, micro- and nano-scale electromechanical system, was introduced to attack, integrate and solve a great variety of emerging problems. Mechatronic systems, as shown in Fig. 9-10, can be classified as(1)conventional mechatronic systems, (2) microeletromechanical-micromechatronic system (MEMS), and (3) nanoelectromechanical-nanomechatronic systems (HEMS). The operational principles and basic fundamentals of conventional mechatronic systems and MEMS are the same, while HEMS are studied using different concepts and theories. In particular, the designer applies the classical mechanics and electromagnetics to study conventional mechatronic systems and MEMS. Quantum theory and nanoelectromechanics are applied in NEMS. The fundamental theories used to study the effects, processes, and phenomena in conventional, micro- and nano-scale mechatronic systems are illustrated in Fig. 9-10.

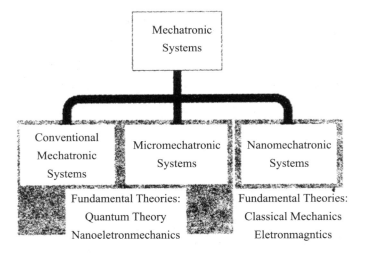

Fig. 9-10 Classification and Fundamental Theories Applied in Mechatronic Systems

Vocabulary Assistant

mechatronics　机电一体化
discipline　学科
synergetic　协同的,合作的
emerge　出现
interdisciplinary　学科间的
benchmark　基准,标准
DSP (Digital Signal Processing)　数字信号处理
optimization　最佳化
multidisciplinary　包括各种学科的
algorithm　算法
virtual　虚拟的
state-of-the-art　最新式的,顶尖水准的
pursue　追求
ultimate　最终的
unify　成为一体
micro-scale　微型
microeletromechanical—micromechatronic　微型电机—机电系统
nanoelectromechanical—nanomechatronic　纳米电机—机电系统
quantum　分配量,额,量

coin　造词
mechanics　机械学
fusion　融合
unattainable　达不到的
paradigm　范例
comprehensive　综合的

prototype　创造原型
system architecture　系统结构
visualization　可视化

downgrade　使降低,使降级
consensus　一致
cornerstone　基石
nano-scale　纳米规模

Over to you:

1) What does the word mechatronics indicate?

2) In mechatronics systems design, what is the most challenging problem?

3) What are the ultimate goals of the mechatronic paradigm?

Vocabulary

A

3-D animation		3D 动画设计	unit 4
abnormal	adj.	异常的,反常的	unit 2
abnormal glow		反常辉光(放电)	unit 7
AC motor		交流电动机	unit 8
access	v.&n.	接入;存取(计算机文件)	unit 4
accessible	adj.	可得到的,易接近的	unit 7
accomplish	v.	实现,完成,贯彻	unit 2
account for		占	unit 9
accordance	n.	相符,一致	unit 6
acquisition	n.	获取	unit 5
activate	v.	(使)起作用	unit 2
actuate	v.	开动(机械等);驱使	unit 9
actuator	n.	传动机构,执行机构,驱动器,传动装置	unit 9
adapter	n.	适配器	unit 4
adjacent	adj.	相邻的,接近的	unit 6
adjustable	adj.	可调整的,可控制的	unit 8
aerospace	n.	航空,航天	unit 2
after-conversion		转变后	unit 5
aid	n.&v.	帮助,辅助	unit 4
alarm	n.	报警器	unit 6
algorithm	n.	算法	unit 9
alignment	n.	并列,定心,定向,定位,校正,调整	unit 3
allowance	n.	公差;容许	unit 2
alloy	v.	合金	unit 2
alnico magnet		铝镍钴磁钢	unit 8
alternating	adj.	交流(电的)	unit 2
altitude	n.	海拔,标高	unit 8
aluminum	n,	铝	unit 2
ambient	adj.	环境的,外界的	unit 8
amenable	adj.	受……影响的	unit 7
amorphous	adj.	无定形的	unit 2
amperage	n.	安培数	unit 2
amplification	n.	扩大,放大,增幅	unit 6

amplificatory	adj.	放大的	unit 7
amplifier	n.	放大器,扩音机	unit 8
amplify	v.	扩大,放大	unit 6
analog	n.	类似物,相似体,模拟	unit 6
analog circuit		模拟电路	unit 6
analogy	n.	类似,模拟	unit 9
analytic	adj.	分析的	unit 1
anatomy	n.	结构,剖析,解剖	unit 9
angle	n.	角度	unit 3
annealing	n.	退火	unit 2
annular	adj.	环状的	unit 8
antifreezing	adj.	防冻的	unit 8
antistatic	adj.	抗静电的	unit 8
apparatus	n.	仪器,装置	unit 8
application	n.	应用	unit 1
approach	n.	方法,途经	unit 5
approximately	adv.	大约	unit 2
apron	n.	溜板,挡板,皮圈	unit 3
arc discharge		电弧放电	unit 7
arc welding		电弧焊	unit 2
armature	n.	电枢	unit 8
articulated	adj.	铰接的,铰链的	unit 9
articulated robot		关节型机器人	unit 9
artificial intelligence		人工智能	unit 2
as a rule		通常	unit 8
aspheric	adj.	非球面的	unit 7
assemble	v.	装配	unit 1
assembling operation		组装业务	unit 9
assembly	n.&v.	集合,装配;组件,集合,装配	unit 3
atom	n.	原子	unit 2
atomic	adj.	原子的,微粒子的	unit 2
attach	v.	连接	unit 7
attribute	n.	属性,品质	unit 9
audio	n.& adj.	音频;音频的	unit 6
austenitic stainless steel		奥氏体不锈钢	unit 2
automate	v.	使……自动化	unit 2
automatic processing		自动化加工	unit 4
automatically	adv.	自动地	unit 2
automation	n.	自动化,自动控制	unit 4
automobile	n.	汽车	unit 2
automotive	adj.	汽车的	unit 1
automotive industry		汽车工业,汽车制造业	unit 1

| axis | *n.* | 轴 | unit 3 |
| axle | *n.* | 轮轴,车轴 | unit 8 |

B

bar	*n.*	条,棒,杆	unit 2
battery	*n.*	电池,蓄电池,电池组	unit 6
bed	*n.*	(机床)床身	unit 3
benchmark	*n.*	基准,标准	unit 9
beverage	*n.*	饮料	unit 2
blanking	*n.*	冲裁,切料,下料	unit 2
block		滑车	unit 1
blow	*n.&v.*	击打	unit 2
blower	*n.*	鼓风机	unit 8
blowing	*n.*	吹炼	unit 2
blown fuse		熔断的保险丝	unit 6
booming	*adj.*	急速发展的	unit 9
boring	*n.*	钻孔	unit 3
boring machine		镗床	unit 3
bracket	*n.*	支架,悬臂	unit 7
brass	*n.*	黄铜	unit 2
breadboard	*n.*	电路试验板	unit 6
brittle	*adj.*	易碎的	unit 2
broaching machine	*n.*	拉床,铰孔床	unit 3
bulk	*n.*	体积;批量	unit 2
burr	*n.*	(金属切削或成型留下的)毛口,毛边	unit 7
burst	*v.*	爆裂,爆发	unit 6
bushing	*n.*	[机]轴衬;[电工]套管	unit 1

C

CIMS (computer integrated manufacturing system)		计算机集成制造系统	unit 4
calculation	*n.*	计算;考虑	unit 1
cam	*n.*	凸轮	unit 1
capability	*n.*	能力	unit 9
capable	*adj.*	有能力的,有可能的	unit 1
capacitor	*n.*	电容器	unit 6
capacity	*n.*	能力;容量;电容	unit 9
carbide	*n.*	碳化物,电石;硬质合金	unit 8
cartesian robot		直角坐标机器人	unit 9
casing	*n.*	机壳	unit 8

cast	n.	铸件,铸型	unit 2
casting	n.	铸件,铸造	unit 2
castiron		铸铁	unit 7
category	n.	种类,范畴	unit 9
cavity	n.	洞,空穴,模腔	unit 2
cellular phone		便携式电话	unit 6
ceramic mould		陶瓷铸型	unit 2
ceramics	n.	陶瓷,制陶术	unit 2
chain	n.	链条,电路	unit 1
characteristic	n.	特征,特性,性能;	unit 9
		形成缺口,碎裂	unit 7
chip	n.	芯片,切屑	unit 6
chromium	n.	铬	unit 2
chuck	n.	卡盘,夹具	unit 3
circuitry	n.	电路图,电路,线路	unit 6
circular	adj. & n.	循环的,圆形的;传单,通报	unit 4
clamp	v.	夹钳,夹具	unit 3
classification	n.	分类,分级	unit 1
cleat	n.	夹板,楔子	unit 3
closed circuit		闭合电路	unit 6
closed loop control system		闭合环路式控制系统,闭环控制系统	unit 8
close-loop		闭环,封闭式回路	unit 9
coarse	adj.	(表面或质地)粗糙的	unit 7
cobalt	n.	钴	unit 2
cockpit	n.	驾驶员座舱	unit 9
coil	n.	线圈,卷材	unit 8
coin	v.	造词	unit 9
coincident	adj.	一致的,符合的	unit 9
combination	n.	结合,联合	unit 1
combine	v.	(使)联合	unit 1
combine harvester		联合收割机	unit 1
command signal		指令信号,控制信号	unit 8
communication	n.	通信	unit 6
commutation	n.	交换	unit 9
commutator	n.	整流器,换向器	unit 8
compacted	adj.	紧密的,紧凑的	unit 2
companionship	n.	交谊,友谊	unit 9
comparison	n.	比较	unit 5
compatible	adj.	相兼容的,能共处的,可并立的	unit 5
competitive	adj.	竞争的,有竞争力的	unit 9
completion	n.	完成	unit 6
complex	adj.	复杂的	unit 1

compliance	n.	依从;柔度	unit 9
component	n.	成分;元件,部件	unit 1
compound	adj.&n.	复合的;混合物	unit 2
compound rest		复式刀架,(车床的)小刀架	unit 3
compound-wound		复式励磁的	unit 2
comprehensive	adj.	综合的	unit 9
compression mould		冲压模,压塑模	unit 2
compressive	adj.	有压缩力的,压缩的	unit 2
compressor	n.	压缩机	unit 8
compulsory	adj.	强制的,强迫的	unit 3
computation	n.	计算,估计	unit 1
computer integrated manufacturing		计算机集成制造	unit 5
computer telecommunication network		计算机通信网络	unit 5
concurrent	adj.	协作的,并发的	unit 9
conductive	adj.	导电的,传导性的	unit 6
conductor	n.	导线,导体	unit 6
configuration	n.	构造,结构;外形;配置	unit 9
conical	adj.	圆锥形的,圆锥的	unit 3
conjunctive	adj.	相关的,相关联的	unit 4
conquer		占领	unit 9
consensus	n.	一致	unit 9
conservative	adj.	保守的;永恒的	unit 9
consist	v.	由……组成	unit 4
consistency	n.	一致,相符	unit 2
consist of		包含,包括	unit 4
constant	adj.	不变的,持续的	unit 2
	n.	常数,恒量	unit 6
constant speed		恒速	unit 8
constraint	n.	限制	unit 4
construction	n.	建筑	unit 2
contact small plate		接触片	unit 8
continuously	adv.	连续地	unit 9
contour	n.	轮廓(等高线,周线,电路,概要)	unit 7
control equipment		控制装置	unit 8
controller	n.	控制器	unit 8
conversion	n.	转换	unit 6
convert	v.	使转变,使……改变	unit 1
conveyance	n.	输送,传导	unit 5
conveyor	n.	输送机	unit 4
convince	v.	说明	unit 5
coolant	n.	冷冻剂	unit 4
coordinate	n.	坐标	unit 9

coordination	n.	协作,协调	unit 4
copper	n.	紫铜,铜	unit 2
copper-tungsten	n.	钨铜合金	unit 7
core	n.	核心,铁心	unit 4
cornerstone	n.	基石	unit 9
correction	v.	校正	unit 2
corresponding	adj.	相应的,相当的	unit 2
corrosion	n.	侵蚀	unit 2
cost-effectiveness	n.	成本效益	unit 5
cotter pin		开口销	unit 1
counter	n.	计算器,计数器	unit 6
counting	n.	计算	unit 6
crack	n.	破裂,崩裂	unit 2
craftwork design		工艺设计	unit 5
crane	n.	起重机	unit 8
crater	n.	凹陷处,焊口	unit 7
crep sole		绉胶鞋底	unit 6
critical strain		临界应变	unit 2
critical zone		临界区域	unit 2
crystalline	adj.	水晶的	unit 2
crystal face		晶面	unit 7
crystal orientation		晶体取向	unit 7
cubic	adj.	立方体的	unit 4
cultural problem		文化问题	unit 5
current	n.	电流,水流,气流	unit 7
cutter	n.	刀具	unit 3
cutting tool		切割具,刀具	unit 3
cycle	n.	周期,循环	unit 5
cylindrical	adj.	圆柱的	unit 3
cylindrical coil		圆柱形线圈	unit 8
cylindrical robot		圆柱坐标机器人	unit 9
Czech	adj.	捷克人(的)	unit 9

D

damping	n.	阻尼	unit 8
data	n.	数据	unit 5
database	n.	数据库	unit 4
DC motor		直流电动机	unit 8
deburr	v.	去除,毛刺,修边	unit 7
de-energize	v.	断路	unit 6
defectively	adv.	有缺陷地,不完全地	unit 6

density	n.	密度	unit 2
deposit	v.	堆积,存放	unit 6
depositing	n.	入库;沉淀;镀层	unit 2
diagram	n.	图表	unit 3
diameter	n.	直径	unit 3
die	n.	钢型,冲模	unit 2
die-casting mould		压铸模	unit 2
differential		差速器	unit 1
differential circuit		微分电路	unit 6
digital	adj.	数字的	unit 2
digital circuit		数字电路	unit 6
digital simulation model		数字仿真模型	unit 4
digitizer	n.	数字化转换器	unit 4
dimension	n.	大小;体积	unit 4
dimensional	adj.	维(数)的,维度的	unit 4
diode	n.	二极管	unit 6
dip	v.	浸,蘸	unit 7
direct current	n.	直流电	unit 2
directive wheel		导向轮	unit 7
discipline	n.	学科	unit 9
discrete	v.	分离的,离散的	unit 4
dismantle	v.	拆除	unit 1
dispate	v.	消散,消失	unit 7
dispersive	adj.	分散的,散布的	unit 6
display system		系统显示器	unit 4
dissipation	n.	消散,分散	unit 6
distortion	n.	扭曲,变形	unit 7
distribution	n.	分配;物流;销售;分布	unit 5
DNC (direct numerical control)		直接数字控制	unit 4
double-suction type		双吸型	unit 6
downgrade	v.	使降低,使降级	unit 9
downloadable	adj.	可下载的	unit 7
drafting	n.	起草	unit 1
drawback	n.	缺点	unit 2
drawing die		拉模	unit 2
drill	v.	钻孔	unit 9
drilling machine		钻床,钻孔机	unit 3
drive	n.&v.	驱动	unit 9
dry cell		干电池	unit 6
DSP (Digital Signal Processing)		数字信号处理	unit 9
ductility	n.	延展性,柔韧性	unit 2
duplicate	adj.	复制的,副本的	unit 2

	v.	复制	unit 3
durability	*n.*	耐久性,耐用	unit 2
durable	*adj.*	耐用的	

E

earth	*v.*	将(电线或其他导体)接地	unit 6
edge	*v.*	镶边,飞边	unit 2
edge radius		刃口半径	unit 7
effector	*n.*	受动器,效应器,操纵装置	unit 9
ejection	*n.*	排放	unit 4
elbow	*n.*	肘,弯管	unit 9
EDM (electrical discharge machining)		电火花加工	unit 7
electrical motor		电动机	unit 9
electrode	*n.*	电极;焊条	unit 4
electromagnetic	*adj.*	电磁铁的,电磁体的	unit 8
electromagnetism	*n.*	电磁,电磁学	unit 8
electronic	*adj.*	电子的	unit 4
electronics	*n.*	电子学(的应用)	unit 2
electronic organ		电子琴	unit 6
electrotechnical	*adj.*	电工技术的	unit 6
elevate	*v.*	提高,升高	unit 2
eliminate	*v.*	消除,剔除	unit 7
emerge	*v.*	出现	unit 9
emergency	*n.*	突发事件	unit 3
emulation	*n.*	仿真,模拟	unit 4
encoder	*n.*	编码器	unit 8
encompass	*v.*	包含	unit4
end effector		末端执行器	unit 9
end shield		端盖	unit 8
energize	*v.*	使电流入	unit 6
energy plant		能源车间	unit 9
engine lathe		普通车床	unit 3
engineering	*n.*	工程(学)	unit 2
engineering design		工程设计	unit 5
enterprise	*n.*	企业	unit 5
entrenched	*adj.*	确立的	unit 9
envelop	*n.*	外壳,包皮	unit 2
equation	*n.*	相等,平衡,方程式	unit 1
estimation	*n.*	判断	unit 6
etch	*v.*	蚀刻	unit 6
exceed	*v.*	超过	unit 8

excitation	*n.*	励磁	unit 8
expansion	*n.*	膨胀	unit 2
expertise	*n.*	专业技术	unit 4
expose	*v.*	暴露	unit 6
external	*n.*	外部	unit 4
external magnetic field		外磁场	unit 8
extrude	*v.*	挤压	unit 2
extrusion	*n.*	挤压	unit 2
extrusion mould		挤压成型模	unit 2

F

fabrication	*n.*	编造	unit 4
facility	*n.*	设施	unit 5
factory mine		工矿企业	unit 8
fall into		分成	unit 5
fan	*n.*	电风扇	unit 8
fan-cooled three phase squirrel-cage asynchronous motor		扇冷式三相鼠笼异步电动机	unit 8
fashion	*v.*	将……做成形态,设计,制造	unit 9
fasten	*v.*	结牢	unit 1
fatigue	*adj.*	劳累	unit 9
feasible	*adj.*	可行的	unit 3
feed	*v.*	进刀,走刀	unit 3
feedback	*n.*	回应,反馈	unit 5
feed-rod	*n.*	进刀杆	unit 3
feed selector		进刀选择器	unit 3
ferrite	*n.*	铁氧体,铁素体,铁酸盐	unit 8
fettle	*v.*	修整,修补	unit 9
field coil		励磁线圈	unit 8
field magnet		场磁体	unit 8
filament	*n.*	细丝,灯丝	unit 6
file	*n.*	锉	unit 3
file manipulation of tool position		加工刀位文件	unit 4
film	*n.*	膜体	unit 6
financial	*adj.*	财政的,金融的	
finish machining		精加工	unit 7
finished product		成品	unit 1
finite-element method		有限元法	unit 2
fixed-sequence robot		固定顺序机器人	unit 9
fixture	*n.*	夹具	unit 4
flame	*n.*	火焰	unit 2

flange	n.	凸缘	unit 8
flashbulb	n.	[摄]闪光灯泡,镁光灯	unit 6
flaw	n.	裂纹	unit 2
flexible manufacturing system (FMS)		柔性加工系统	unit 4
flow stress		流动应力	unit 2
fluid bath		液槽	unit 7
flush away		冲去	unit 7
foil	n.	箔,金属薄片	unit 2
follower	n.	从动轮	unit 1
forging	n.	锻造,锻件	unit 2
forging die		锻模	unit 2
formulation		明确表达	unit 5
fracture	n.	断裂	unit 2
fracture toughness		断裂韧度	unit 2
fragile	adj.	易碎的,脆的	unit 7
frequency	n.	频率	unit 7
frequency synthesizer		频率合成器	unit 6
functional block		功能块	unit 6
fuse	n.	导火线,保险丝	unit 6
fused silica		熔融石英	unit 7
fusion	n.	融合	unit 9

G

gagging	n.	成型	unit 4
gantry	n.	构台,桶架	unit 9
gantry robot		拱架机器人	unit 9
gap	n.	凹口,开口,间隙	unit 3
gear	n.	齿轮	unit 1
gearbox	n.	变速箱	unit 3
general-purpose	adj.	多种用途的	unit 9
generate	v.	生成,产生	unit 2
generative	adj.	生成的,生产的	unit 4
generator	n.	发电机,发生器	unit 2
geometric model		几何模型	unit 4
geometry	n.	几何学	unit 2
grain growth		晶粒生长	unit 2
grain size		结晶粒度(大小)	unit 2
graphic	n.&adj.	制图,制图法;图表的,图画的	unit 4
graphite	n.	石墨	unit 7
grid	n.	格子,网格,点阵	unit 6
grind	v.	磨,碾	unit 7

grinder	*n.*	磨床,研磨机	unit 7
grinding machine		磨床	unit 3
grinding paste		磨削用冷却剂	
grinding plate		砂盘	unit 7
grit	*n.*	粗砂,沙砾	unit 3
groove	*n.*	槽纹,刻槽	unit 1
guarantee	*v.&n.*	保证	unit 5

H

hair crack		细裂纹	unit 7
hammer	*v.*	锤击,捶打	unit 2
hand tool		手工具	unit 3
hardened tool-steel		淬硬工具钢	unit 7
harness	*v.*	固定,控制	unit 8
hazard	*n.*	危险,冒险	unit 2
headstock	*n.*	床头箱,主轴箱	unit 3
headstock assembly		主轴箱组件	unit 3
heal	*v.*	治愈,结束,恢复	unit 2
heat treatment		热处理	unit 2
heavy-duty		耐受力强的,重型的	unit 3
helix	*n.*	螺旋,螺旋状物	unit 1
hexagonal	*adj.*	六角形的	unit 3
high reliability		高可靠性	unit 6
high resistance		高阻的	unit 8
high impact	*n.*	高冲击	unit 8
high-intensity		高强度的	unit 8
high-speed drive		高速驱动	unit 8
hoist	*n.*	卷扬机,升降机	unit 8
holder	*n.*	固定器,支持物,支架	unit 3
homogeneity	*n.*	同质,均匀性	unit 2
hot forging		热模锻	unit 2
hp (horse power)		马力	unit 8
humanoid	*adj.*	具有人的特点的	unit 9
hydraulic	*adj.*	水力的,液压的	unit 9
hydraulic press		水压机,液压机	unit 2
hydraulic system		液压系统	unit 9

I

identify	*v.*	辨别,识别	unit 4
impact	*v.&n.*	冲压,冲击	unit 2

imparting	adj.	传授的,告知的	unit 2
implementation	n.	执行,实现	unit 5
imply	v.	暗示	unit 2
impression	n.	压痕	unit 2
in quantity		大量的	unit 2
incompatibility	n.	不相容	unit 5
incorporate	adj.&v.	合并的,一体化的;合并,结合	unit 4
incredible	adj.	难以置信的	unit 4
induce	v.	产生,带来	unit 8
inductance	n.	感应系数,自感应	unit 8
induction motor		感应电动机	unit 8
inertia	n.	惯性,惯量	unit 8
information	n.	信息	unit 5
ingot	n.	锭铁,工业纯铁	unit 2
inherent	adj.	固有的,内在的	unit 2
inhibit	v.	抑制,禁止	unit 5
injection mould		注塑模	unit 2
innovation	n.	创新,革新	unit 5
inorganic	adj.	无机的	unit 2
input	n. v.	输入	unit 6
inspection	n.	检查,视察	unit 2
install	v.	安置	unit 9
installation	n.	安装	unit 6
instant	n.	瞬间	unit 8
instantaneous	adj.	瞬间的,即时的	unit 5
insulating base		绝缘底板	unit 7
insulation class		绝缘等级	unit 8
insulator	n.	绝缘体	unit 7
integral	adj.	构成整体所必需的,完整的	unit 8
integrated circuit		集成电路	unit 6
integration	n.	集合,综合,集成	unit 6
integral circuit		积分电路	unit 6
intelligent	adj.	智能的	unit 9
intelligent robot		智能机器人	unit 9
intelligible	adj.	可理解的	unit 1
interaction	n.	相互作用,相互影响	unit 8
interactive	adj.	交互式的	unit 4
interconnect	v.	使互相连接	unit 6
interdisciplinary	adj.	学科间的	unit 9
intermetallic	adj.	金属间的	unit 2
intermittent	adj.	间歇的,断断续续的	unit 8
internal configuration		内部构形	unit 8

interpolation	*n.*	内插,内插法,窜改	unit 4
intricate	*adj.*	复杂的	unit 2
intrude	*v.*	入侵	unit 8
invaluable	*adj.*	无价的,非常宝贵的	unit 1
involve	*n.*	与……有关联,涉及	unit 1
iron filing		铁屑	unit 8

J

jaw	*n.*	钳夹	unit 3
jig	*n.*	机床夹具	unit 4
joint	*n.*	关节,结合点	unit 9

K

| keyboard | *n.* | 键盘 | unit 4 |
| knurl | *adj.* | 压花的,压边的 | unit 3 |

L

label	*n.*	标签	unit 6
latent parameter		特征参数	unit 4
lathe	*n.*	车床	unit 3
lead	*n.*	铅	unit 2
lead time		研制周期,交付周期	unit 4
LED (Light Emitting Diode)		发光二极管	unit 6
lethal	*adj.*	致命的	unit 3
lever	*n.*	杆,杠杆	unit 1
life expectancy		平均寿命,预期寿命	unit 1
light beam		线偏振光束,光束	unit 6
linear IC		模(线性)集成电路	unit 6
linear	*adj.*	线性的,直线的,线状的	unit 6
linearity	*n.*	线性,直线性	unit 8
liquid	*adj.*	液体的,流动的	unit 2
loading	*n.*	装载	unit 9
local storage		局部存储器	unit 4
logic	*n.*	逻辑(学)	unit 6
longitudinal	*adj.*	纵向的	unit 3
loop	*n.*	环,圈,弯曲部分	unit 1
low-carbon		低碳(的),含碳量低(的)	unit 2
low cost		低成本	unit 4
low-inertia	*adj.*	低惯性的	unit 8

| lubricant | n. | 润滑剂,润滑油 | unit 3 |
| lubricate | v. | 使润滑 | unit 3 |

M

machine loading		机器装载,机器负荷	unit 9
machine tool		机床,工具机	unit 4
machining operation		机械作业	unit 3
magnet	n.	磁铁	unit 8
magnetic force		磁力	unit 8
magnetic mould		磁铁成型模	unit 2
magnify	v.	扩大,夸大	unit 6
mains supply		干线供电,交流电源	unit 2
maintenance	n.	维护	unit 3
management	n.	管理,操纵	unit 4
management administration		经营管理	unit 5
management policy		经营方针,管理策略	unit 5
manipulate	n.	操作	unit 4
manipulator	n.	操作者,操纵器	unit 9
manual data input		手动数据输入	unit 4
manufacturing automation		制造自动化	unit 5
manufacturing process		制作过程	unit 4
manufacturing quality		加工质量	unit 4
manufacturing tolerance		制造公差	unit 1
market competition		市场竞争	unit 5
materials handling		物料输送;原材料处理	unit 5
material processing		材料加工	unit 9
mechanical press		机械冲床	unit 2
mechanical property		机械性能,力学性质	unit 2
mechanics	n.	机械学	unit 9
mechatronics	n.	机电一体化	unit 9
menace	n.	威胁,恐吓	unit 3
mesh	n.&v.	网格,网状结构;协调;啮合	unit 1
metal die		金属压型	unit 2
metal-cutting machine		金属切削机械	unit 8
metallurgy	n.	冶金,冶金学	unit 2
microeletromechanical- micromechatronic	n.	微型电机—机电系统	unit 9
micrometer	n.	微米,百万分之一米(长度单位,符号为μm)	unit 6
		千分尺	unit 3
micro-scale	adj.	微型的	
mill	n.	轧钢机	unit 2

milling machine		铣床	unit 3
minimum	*n.*	最小值	unit 3
mining machine		采掘机,矿山机械	unit 8
minor	*adj.*	次要的	unit 9
minus	*adj.*	负的	unit 6
mobil	*adj.*	可移动的	unit 9
mixer	*n.*	混合机,搅拌机	unit 8
model	*n.*	模型	unit 4
mold	*n. & v.*	模子,铸型;模压	unit 2
molten	*adj.*	融化的	unit 2
molten state		熔融状态	unit 2
molybdenum filament		钼丝	unit 7
monitoring and control		计算机监控	unit 4
monolithic	*adj.*	整体式的	unit 8
	n.	单块集成电路	unit 6
monolithic technique		单块技术	unit 6
motion	*n.*	运动,动作	unit 9
motor-generator set		电动发电机设备	unit 8
mould/mold	*n.&v.*	模具,模子;使成型,模型	unit 2
moulding	*n.*	成型,造型	unit 2
mount	*v.*	安装	unit 3
movable core		动铁芯	unit 8
moving-coil		动圈式	unit 8
msec=megasecond		兆秒	unit 8
multi-function manipulator		多功能操纵器	unit 9
multidisciplinary	*adj.*	包括各种学科的	unit 9
multimedia	*n.*	多媒体	unit 4
multipurpose	*adj.*	多种用途的	unit 9
multi-service	*n.*	多服务	unit 4
multi-spindle automatic lathe		多轴自动车床	unit 3
mutational	*adj.*	变化的,转变的,突变的	unit 6

N

nanoelectromechanical- nanomechatronic	*n.*	纳米电机—机电系统	unit 9
nanometer	*n.*	纳米	unit 7
nano-scale		纳米规模	unit 9
negative terminal		负极接线柱	unit 6
nickel	*n.*	镍	unit 2
non-ferrous	*adj.*	有色的,非铁或钢的	unit 2

nuisance	n.	令人讨厌的东西	unit 1
numerical	adj.	数字的	unit 5
numerically controlled robot		数控机器人	unit 9

O

object	n.	目标	unit 6
occur	v.	发生,出现	unit 2
off-line		脱机,离线	unit 9
open circuit		断路,开路	unit 6
open-loop		开环,开放式回路	unit 9
operator	n.	操作者	unit 9
opposite	adj.	对面的;对立的	unit 2
optical	adj.	光学的	unit 7
optics	n.	光学	unit 7
optimal	adj.	最优的,最佳的	unit 4
optimization	n.	最佳化	unit 9
order component		序分量	unit 4
orient	v.	使适应	unit 2
oscillation	n.	摆动,振动	unit 1
output	n.&v.	产量,输出(量)	unit 6
overall	adj.	全部的,全体的	unit 1

P

packaging	n.	包装	unit 5
palletizing	n.	装运	unit 9
parabolic	adj.	抛物线的	unit 4
parabolic wave		抛物线形波	unit 6
paradigm	n.	范例	unit 9
parallel	adj.	平行的,相同的	unit 9
parallel circuit		并联电路	unit 6
parameter	n.	参数	unit 4
part	n.	零件,部件	unit 4
particularity	n.	特性,特质	unit 1
patent	n.	专利	unit 9
periodic table		(元素)周期表	unit 2
permanent	adj.	永久的	unit 1
permanent magnet		永磁体	unit 8
perpendicular	adj.	垂直的,正交的	unit 8
petrol engine		汽油(发动)机	unit 2
photocell	n.	光电池,光电管	unit 6

physics	*n.*	物理过程,物理现象	unit 6
piezoelectric	*adj.*	压电的	unit 2
pin	*n.*	栓,管(脚)	unit 6
pivot point		支点	unit 1
pixel	*n.*	像素	unit 4
plane	*adj.*	平面的	unit 1
planer	*n.*	龙门刨床	unit 3
plastic	*adj.*	塑造的,有可塑性的	unit 2
plate	*n.*	中厚(钢板)	unit 2
platform	*n.*	平台,月台,讲台	unit 9
player	*n.*	影碟机	unit 6
playback robot		再生式机器人	unit 9
plier	*n.*	钳子,镊子	unit 3
plotter	*n.*	数据自动描绘器,绘图机	unit 4
plus	*adj.*	正的	unit 6
pneumatic	*adj.*	气力的	unit 9
polar	*adj.*	极性的	unit 9
polar coordinate		极坐标	unit 9
polarity	*n.*	磁性引力,极性	unit 8
polar robot		极坐标机器人	unit 9
polish	*v.*	磨光	
polishing	*n.*	磨光,磨料	unit 7
poly-phase		多相,多相的	unit 8
polymer	*n.*	聚合体	unit 2
porosity	*n.*	多孔性	unit 2
portable	*adj.*	轻便的,便携式	unit 2
portion	*n.*	部分	unit 7
positive	*adj.*	正的	unit 6
possess	*v.*	占有,拥有	unit 9
powder	*n.*	粉末	unit 2
powder metallurgical mould		粉末冶金模	unit 2
practical	*adj.*	实际的	unit 4
precaution	*n.*	预防,预防措施	unit 3
precision	*n.*	精确,精密度	unit 9
pre-conversion		转变前	unit 5
predetermine	*v.*	事先安排	unit 1
preliminary	*adj.*	初步的	unit 4
prime	*adj.*	首要的,重要的	unit 1
principal	*adj.*	主要的	unit 2
printed circuit		印刷电路	unit 8
prior to		在……之前	unit 3
prismatic	*adj.*	棱镜的,柱状的	unit 7

process flow		工艺流程	unit 4
processing parameter		工艺参数	unit 4
procurement management		采购管理	unit 5
product data management		产品数据管理	unit 4
product of advanced technology		高新技术产品	unit 5
profile	n.	断面,剖面	unit 2
program block		程序块	unit 2
program file		程序文件	unit 7
programmed motion profile		程序运动轮廓	unit 8
prohibitive	adj.	禁止的	unit 3
proportional	adj.	成比例的	unit 2
protection class		防护等级	unit 8
protocol	n.	协议	unit 4
prototype	v.	创造原型	unit 9
proto-typing	n.	初始制模	unit 2
pulley	n.	滑车,滑轮	unit 1
pulsating	adj.	脉动的;极为兴奋的	unit 7
pulse power source		脉冲电源	unit 7
pursue	v.	追求	unit 9
punch	v.	冲孔,打孔	unit 2
punched tape		穿孔纸带	unit 9

Q

quantum	n.	分配量,额,量	
quill	n.	套筒轴	unit 3
quoted price		报价	unit 5

R

radial	adj.	放射(式)的,辐射(式)的	unit 8
radius	n.	半径,辐射线	unit 3
ram EDM		介质放电加工	unit 7
rated frequency		额定频率	unit 8
rated voltage		额定电压	unit 8
ratio	n.	比率	unit 9
rational	adj.	理性的,合理的	unit 1
rattle	n.	一连串短促而尖锐的声音	unit 1
raw material		原材料	unit 5
ream	v.	铰孔,扩孔	unit 3
reconfigure	v.	改造,更换部件	unit 4
recrystallize	v.	再结晶	unit 2

recrystallization	*n.*	再结晶	unit 2
rectangular	*adj.*	矩形的,成直角的	unit 9
rectification	*n.*	纠正,整顿	unit 2
rectifier	*n.*	整流器	unit 8
refer	*v.*	涉及	unit 9
refer to		参考,涉及,关于,查阅	unit 1
refinement	*n.*	细化,精练	unit 2
regenerative	*adj.*	再生的	unit 6
relative humidity		相对湿度	unit 8
relatively	*adj.*	相对地	unit 2
release	*v.*	释放	unit 6
reliability	*n.*	可靠性	unit 1
reliably	*adv.*	可靠地	unit 9
relieve	*v.*	解除	unit 2
remotely	*adv.*	远程地	unit 8
remote control		遥控	unit 6
repetitive	*adj.*	反复的	unit 9
reprogram	*v.*	改编,程序重调	unit 9
resistance	*n.*	抵抗力	unit 2
resistor	*n.*	电阻器	unit 6
resolver	*n.*	分解器	unit 8
retrieval	*n.*	检索	unit 4
reverse	*v.*	颠倒,倒转	unit 1
revolve	*v.*	旋转	unit 3
rigid	*adj.*	刚硬的,坚硬的	unit 1
robotics	*n.*	机器人技术	unit 5
roller	*n.*	辊子,滚筒	unit 2
rolling	*n.*	辗,轧	unit 2
rotary	*n.*	旋转,循环	unit 9
rotate	*v.*	旋转	unit 1
rotating shaft		转轴	unit 1
rotation	*n.*	旋转	unit 3
rotational motion		回转运动	unit 8
rotational speed		转动速度,周围速度	unit 8
rotor	*n.*	转子,回转轴,转动体	unit 8
rough machining		粗加工	unit 7
rough part		毛坯,半成品	unit 4
ruggedness	*n.*	坚固性	unit 8
run-off	*n.*	流出口,流放口	unit 4

S

saddle	n.	鞍座,管托,座架	unit 3
sales management		销售管理	unit 5
sapphire	n.	蓝宝石	unit 7
sawtooth	n.	锯齿	unit 6
SCARA (selective compliance assembly robot arm)		平面关节型机器人	unit 9
scheduling	n.	安排,调度	unit 4
scrap	n.	废料	unit 9
screw	n.	螺丝钉	unit 1
sea level		海平面	unit 8
sealant	n.	密封剂	unit 9
secondary	adj. & n.	次级的;次级电路	unit 2
secure	adj.	牢固的	unit 8
semiautomatic	adj.	半自动的	unit 2
semiconductor	n.	半导体	unit 6
semi-permanent		半永久的,暂时的	unit 1
sensor	n.	传感器	unit 9
sensitive	adj.	敏感的,灵敏的	
sequence	n.	次序,顺序	unit 9
sequential	adj.	连续的,相继的	unit 2
series circuit		串联电路	unit 6
series-parallel		串并联	unit 6
servo motor		伺服电动机	unit 8
servocontrol	n.	伺服控制	unit 9
setup	n.	装备,设备	unit 3
shaft	n.	轴	unit 1
shaft configuration		轴结构	unit 8
shaft resonance		轴共振	unit 8
shaper	n.	牛头刨床	unit 3
shearing machine		剪切机	unit 3
sheet	n.	薄层,板	unit 2
shoulder joint		肩关节	unit 9
short-circuit		短路	unit 6
shrinkage	n.	收缩	unit 2
shut-off	n.	截流,断流	unit 4
signal	n.	信号	unit 4
silicon	n.	硅	unit 6
silver-tungsten	n.	钨银合金	unit 7
simulation	n.	仿真,模拟	unit 5
simulator	n.	模拟器	unit 9

simultaneously	*adv.*	同时地	unit 6
sine wave		正弦波	unit 8
single-phase		单相	unit 8
single-spindle automatic lathe		单轴自动车床	unit 3
single-suction type		单吸型	unit 6
sintering	*n.*	烧结	unit 2
site	*n.*	场所,地点	unit 2
sketch	*n.*	草图,设计图	unit 1
slide	*v.&n.*	滑动;滑板	unit 3
smooth	*v.*	使光滑,使平坦	unit 2
software	*n.*	软件	unit 4
solid	*adj.&n.*	坚固的,固体的;固体,凝固	unit 2
solid model		实体模型	unit 4
solidify	*v.*	(使)凝固	unit 2
sort	*v.*	整理,分类	unit 9
sorting	*n.*	分类,整理	unit 9
soundness	*n.*	可靠性,稳固	unit 2
spacecraft	*n.*	太空船	unit 6
space quadrature		空间90度相位差	unit 8
sparking	*n.*	冒火星,产生火花	unit 8
spatial linkage		空间联结	unit 1
specify	*v.*	列举,规定	unit 1
speed-torque		转速力矩	unit 8
spherical	*adj.*	球形的	unit 9
spherical robot		球形坐标机器人	unit 9
spindle	*n.*	心轴,定轴	unit 3
splashy object		飞溅物体	unit 8
spray coating		喷涂	unit 9
spray painting		喷漆	unit 9
spring balance		弹簧秤	unit 1
stainless	*adj.*	不锈的	unit 2
stamping	*n.*	冲压,冲压件,模锻	unit 2
standardize	*adj.*	标准的,定型的	unit 1
state-of-the-art		最新式的,顶尖水准的	unit 9
static	*adj.*	静态的,静力的	unit 6
static electricity		静电	unit 6
stationary	*adj.*	静止不动的,不变的	unit 8
statistics	*n.*	统计	unit 9
stator	*n.*	固定子,固定片	unit 8
status quo		现状	unit 5
steam hammer		蒸汽锤	unit 2
steering wheel		方向盘,转向轮	unit 1

step-down		低压,低压的	unit 2
stereo amplifier		立体声扩音器	unit 6
stiffness	*n.*	坚硬,硬变	unit 2
stock	*v.*	存货	unit 9
storage	*n.*	储存	unit 5
storage of file		文档存储	unit 4
straightforward	*adj.*	正直的;简单的,直截了当的	unit 4
strain	*v.*	拉紧	unit 2
	n.	变形,应变	
strain hardening		机械 [加工,应变]硬化	unit 2
strength	*n.*	强度,力量	unit 2
stress	*n.*	应力,压力	unit 1
stress-free	*adj.*	无应力的	unit 7
stretch out		伸展,伸出	unit 8
stride	*n.*	大步,进步	unit 2
strike	*v.*	拉,撞击	unit 2
strip	*n.*	条,带 ,狭长的一块材料	unit 2
stripe	*n.*	条纹 ,细条	unit 6
structural beam		结构横梁	unit 2
structurally-complicated product		结构复杂的产品	unit 5
subassembly	*n.*	部件,组件	unit 1
subdivide	*v.*	细分,再分	unit 2
submerge	*v.*	使浸水,潜入水中,使陷入	unit 7
subroutine	*n.*	子程序	unit 4
subsequent	*adj.*	随后的,后来	unit 2
successively	*adj.*	一个接一个地	unit 7
suited	*adj.*	合适的	unit 2
summarize	*v.*	概述,摘要而言	unit 5
superconductor	*n.*	超导体	unit 2
surface finish		表面加工,表面光洁度	unit 7
swarf	*n.*	金属屑	unit 3
switching	*n.*	开关	unit 6
synchronous motor	*n.*	同步电动机	unit 8
synergetic	*adj.*	协同的,合作的	unit 9
synthesize	*v.*	合成,结合	unit 1
system	*n.*	系统	unit 5
system architecture		系统结构	unit 9

T

tachometer	*n.*	转速计	unit 8
tackle		索具	unit 1

tactile	*adj.*	触觉的	unit 9
tailstock	*n.*	尾架,尾座,顶针座	unit 3
tailstock assembly		尾座组件	unit 3
technical condition		技术条件	unit 8
technics	*n.*	工艺	unit 6
technique	*n.*	技术	unit 4
technology	*n.*	技术,工艺	unit 5
test	*v.*	测试	unit 9
thermal conductivity		导热性	unit 2
thermal insulator		热绝缘体	unit 2
tight tolerance		紧密度公差,精密公差	unit 7
time constant		时间常数	unit 6
titanium	*n.*	钛	unit 6
tolerance	*n.*	公差	unit 1
tool bit		刀头,刀片	unit 3
tool fitting		配件	unit 3
tool path		刀具轨迹	unit 4
tool-post		刀座,刀架	unit 3
torque	*n.*	扭矩,转矩,力矩	unit 1
tooth	*n.*	轮齿	unit 1
tracing and duplicating lathe		仿形车床	unit 3
track	*n.*	轨迹	unit 6
transfer	*v.*	转移	unit 9
transformer	*n.*	变压器	unit 2
transistor	*n.*	晶体管	unit 6
translation	*n.*	移动,变换	unit 1
transmission	*n.*	传播,发送	unit 5
transmit	*v.*	传输,转送	unit 1
transport platform		传输平台	unit 4
triangular	*n.*	三角形	unit 6
trigger	*v.*	引发,触发	unit 6
trimming	*n.*	修饰加工,切边	unit 2
turnaround	*n.*	转变,转盘	unit 7
turret		六角刀架,转塔刀架	unit 3
turret lathe		六角(转塔)车床	unit 3

U

ultimate	*adj.*	最终的	unit 9
unattainable	*adj.*	达不到的	unit 9
uniform motion		均匀(等速)运动	unit 1
uniform torque		不变的扭动力	unit 8

uniformity	n.	同样,一致	unit 1
unify	v.	成为一体	unit 9
universal motor		交直流两用电动机	unit 8
unload	v.	卸载	unit 9
ultra-precision	adj.	超密的	unit 7
update	v.	更新	unit 5
utility	n.	效用	unit 2
utilize	v.	利用	unit 5

V

variable-sequence robot		可变顺序机器人	unit 9
vary	v.	改变	unit 2
velocity	n.	速度,速率	unit 7
velocity command signal		速度控制信号	unit 8
verification	n.	证实,核实	unit 2
verify	v.	查证,核实;(在DOS命令下)打开/关闭在DOS操作期间的写文件校验开关	unit 4
versatile	adj.	多技能的,多才多艺的	unit 3
versatility	n.	多功能性	unit 2
vertical	adj.	立式	unit 3
vibration	n.	颤动,振动	unit 3
vicinity	n.	在周围地区,在附近	unit 6
virtual	adj.	虚拟的	unit 9
visual	adj.	视觉的	unit 9
visualization	n.	可视化	unit 9
voltage	n.	电压,电位差,伏特数	unit 2

W

warehouse	n.	仓库,材料库	unit 9
waveform	n.	波形	unit 6
wear resistance		耐磨性	unit 2
weld	v.	焊接	unit 1
welder	n.	焊机	unit 4
white heat		白热	unit 2
winding	n.	绕组,线圈	unit 8
wire EDM		丝电火花加工	unit 7
wire-cutting		线切削	unit 3
wire-frame model		线框模型	unit 4
wireless	adj.	无线	unit 4
withdraw	v.	缩回,撤出	unit 9

withstand	v.	抵挡, 经受	unit 2
wobble	v.	摇摆	unit 3
work flow		工作流程	unit 4
working procedure		工序	unit 4
workpiece	n.	工件	unit 2
workshop	n.	车间, 工厂	unit 3
wound-rotor induction motor		线绕式转子感应电动机	unit 8
wrist	n.	腕, 机械腕	unit 9

Y

yield	v.	屈从, 屈服; 产出	unit 2

机电英语（第二版）

尊敬的老师：

　　您好！

　　为了方便您更好地使用本教材，获得最佳教学效果，我们特向使用该书作为教材的教师赠送本教材配套参考资料。如有需要，请完整填写"教师联系表"并加盖所在单位系（院）公章，免费向出版社索取。

北京大学出版社

教 师 联 系 表

教材名称	机电英语 (第二版)		
姓名：	性别：	职务：	职称：
E-mail：	联系电话：	邮政编码：	
供职学校：		所在院系：	（章）
学校地址：			
教学科目与年级：		班级人数：	
通信地址：			

　　填写完毕后，请将此表邮寄给我们，我们将为您免费寄送本教材配套资料，谢谢！

北京市海淀区成府路205号
北京大学出版社外语编辑部　李　颖
邮政编码：100871
电子邮箱：evalee1770@sina.com

邮 购 部 电 话：010-62534449
市场营销部电话：010-62750672
外语编辑部电话：010-62754382